Caverns, Cauldrons, and Concealed Creatures

Second Edition

by

Wm. Michael Mott

Hidden Mysteries

TGS Publishers

Caverns, Cauldrons, and Concealed Creatures - Second Edition
by Wm. Michael Mott

Hidden Mysteries

TGS Publishers
22241 Pinedale Lane
Frankston, Texas 75763
903-876-3256
www.HiddenMysteries.com
info@hiddenmysteries.com

Printed and bound in Texas.

The Texas Constitution
Article 1 - BILL OF RIGHTS
Section 8 - FREEDOM OF SPEECH AND PRESS

Every person shall be at liberty to speak, write or publish his opinions on any subject, being responsible for the abuse of that privilege; and no law shall ever be passed curtailing the liberty of speech or of the press.

ISBN 0-9786249-4-7
EAN: 97809786249-4-1

This book is dedicated to all those Seekers after the Truth behind Mysteries, those curious souls who are not deterred by terror, opposition, or ridicule, from connecting the dots... for regardless of how unrelated these strange specks of data may seem, an image of clarity will eventually be revealed to the steady and patient gaze.

Other Books by

Wm. Michael Mott

Pulsifer: A Fable

Land of Ice, A Velvet Knife

TABLE OF CONTENTS

Foreword by Brian J Allan 9
Foreword by Dean Dominic De Lucia 13
Foreward by Jan Lamprecht 17
THROUGH A GLASS, DARKLY 19
THE LONG NIGHT of HISTORY 25
Part I: THE DEEP DWELLERS 29
PREFACE 29
SECTION 1: MYTHICAL (ANCIENT RELIGIOUS) ACCOUNTS 31
1. The East. 31
2. The West. 35
3. The New World. 39
SECTION TWO: LITERARY EXCURSIONS AND REVELATIONS 45
Part One: European 45
FOUNDATIONAL FORMS 45
1. Wagner. 45
2. Dante Alighieri. 46
3. Jules Verne. 47
FAERIE LANDS REVISITED 49
4. George MacDonald. 49
5. J.R.R. Tolkien. 49
Part Two: North American 51
CAVERNS, GROTTOES, & ELDRITCH HORRORS 51
1. A. Merritt. 51
2. H. P. Lovecraft. 52
3. Robert E. Howard. 53
HOLLOW MOONS, HOLLOW WORLDS 55
4. Edgar Rice Burroughs. 55
PERSPECTIVES 59
SECTION THREE: MODERN FORTEAN AND CRYPTOZOOLOGICAL CONNECTIONS 63
PART ONE: CRYPTIDS AND MYSTERY CREATURES 63
1. Lake Monsters and Serpents. 64

2. Hairy Humanoids. 66
3. El Chupacabras and Other "Hybrid" Beings 67
4. Other Forms. 70
PART TWO: RE-INVENTION, PARANOIA, AND DECEPTION 72
1. THE DEGENERATES BELOW. 72
2. THE MASKS MAY CHANGE, THE PLAY'S THE SAME 77
3. INTO THE OUT OF 83
Part II: HIDDEN NEIGHBORS: 91
1. OUR SURFACE FARM, AND PLAYGROUND 91
2. THE MORE THINGS CHANGE, THE MORE THEY STAY THE SAME116
3. HAMBURGERS FOR HELL, AND OTHER NASTY DISHES 129
4. WHAT IN THE WORLD.....? 135
5. HAIR TODAY, GONE TOMORROW 145
6. THE FARMERS IN THE DELVE 161
Part III: ECHOES of the WORLD CAULDRON: 165
LEGENDARY, MYTHIC AND RELIGIOUS BELIEF IN LANDS BENEATH THE EARTH 165
Part IV: DEMONSEED: CHANGELINGS, HALFLINGS, AND HUMAN MONSTERS 197
1. PREFACE 197
2. IN DAYS OF YORE 200
3. UNHOLY LIAISONS 208
4. THE BEAST WITHIN 221
5. WHO, WHAT, AND FROM WHERE? 231
AFTERWORD 235
Part V: EARTH, MOTHER of MYSTERIES 237
1. A HIDDEN BIOSPHERE 238
2. MYTH AND LEGEND 239
3. DRAGON SPOUSES 239
4. WATER-CATTLE 240
5. THE MOHOROVICIC DISCONTINUITY 242
6. THE ANCIENT ENEMY 244
7. ENOCH 246
8. ALIENS OR HYBRIDS? 247

9. COSMIC DISINFORMATION 249
Part VI: ENTITIES, ELECTROMAGNETISM, and EMPs 251
1. AMBASSADORS FROM HELL 252
2. HAIRY, HAIRY, QUITE CONTRARY 256
3. THE PROBLEM OF DENSITY 260
Part VII: The Nephilim Giants 269
Part VIII: The Shape-Shifting Myth and Mystery 277
Part IX: A Cave Fable 283
Part X: THE DECEPTION of the DEPTHS: 287
Humanity and the Ocean of Secrecy 287
1. OUR MYSTERIOUS HOME 287
2. WHERE DRAGONS ROAM 297
3. WATER, WATER, EVERYWHERE 312
4. THE PERPETUAL REBELLION 338
BIBLIOGRAPHY non-fiction sources: 355
BIBLIOGRAPHY of Fiction Sources: 358
IMAGES 359
ABOUT THE AUTHOR 361
Shaver Rock Images 363
The Angry Twins 365
The Beauty Queen 366
The Incan King 367
The Wizard 368
The Whisperers 369
The Little Demon 370
The Face Cluster 370

Foreword
by Brian J Allan

My own journeys into the hinterland of what is possible, or indeed probable, have revealed many occasions where the nature of reality becomes amorphous and one soon comes to understand that what we see is, in many instances, what we are programmed to see and not what is actually there. The investigations conducted by me and recorded in the books I have written allowed me a privileged and unique glimpse behind the thin veil that separates the many possible worlds that, in various degrees of concord, simultaneously coexist with our own. There are entities living on the Earth in the here and now that remain hidden, mainly due to the fact that we either do not wish to, or are unable to, see them. However, in some cases they may invisible simply because they live at a different frequency than our own conventions allow or permit and theirs alone is the talent to adjust their own rate of existence to harmonise with our own.

This of course chimes exactly with observations of scientists both old and new, for example, the late and near legendary icon, Nikola Tesla, when asked if there was a fundamental secret at the heart of the universe reportedly commented, *'resonance, vibration and frequency'*. This quotation is almost precisely mirrored by the theoretical physicist, Professor Michio Kaku, of New York City University, who makes the rather beautiful observation that the universe is like a vast cathedral resonating to the 'music' of vibrating, infinitesimally small, superstrings. These are probably the smallest theoretical particles in existence; in terms of size a superstring is to an atom as an atom is to the universe, they hold groups of other subatomic particles, called quarks, together.

This being the case, then it should come as no surprise that when we listen to a piece of conventional music that it should produce a series of effects, intended or otherwise, in the listener, for we are after all creatures of frequency. We are all emotionally affected by music, either spiritually elevated and joyous or reduced to deep melancholy and even tears, this is amply illustrated by the exceptionally emotive and moving theme tune from a 1970's TV series entitled, *'Harry's Game'*, written by the Irish Celtic band, Clannad. Sadly, the more unscrupulous amongst us, i.e. big business, use music, and even specific musical frequencies, for their own ends, usually designed to exploit a variety of fundamental human emotions. Perhaps it is this basic building block of humanity that is likewise exploited by the creatures coexisting beside us in order to remain invisible.

This leads to a deeply disturbing range of possibilities regarding the nature and location of the entities that have bedevilled humanity throughout the ages. They have been known by a variety of names, Goblins, Trolls, Elves, Leprechauns, Pixies (and depending on location, Piskies, especially in the beautiful, south western English county of Cornwall, which is also traditionally a place of magic and mysticism and, perhaps typically, in former days a site of intensive mining), also the Blue Men, the Sidhe, Boggarts and of course, Fairies. It should be noted that there is also physically much larger subdivision of fairies called '*the faerie*' that are traditionally inimical to humanity and can be overtly hostile and occasionally malignant. These beings have existed in the traditions and race memories of every civilisation throughout history, even in the humourless, stony materialism and atheism practised in the former USSR and currently in China, but in spite of the best efforts of the communist hierarchy the traditions persisted along with a deep rooted religious undercurrent, for like most of the worlds nations, inherently the Russians and the Chinese are deeply spiritual.

The invisible races are often presented in a homogenous, 'one size fits all' package that includes ghosts, ghouls and even extraterrestrials, although their true points of origin are normally obscured in a haze of confusion and speculation. However, one probable source is the subject of, **'Caverns, Cauldrons and Concealed Creatures'**, a superbly researched and very accessible book by Wm. Michael Mott, who ably demonstrates that the pits and crevasses hidden deep below the surface of planet Earth might just hold the answer to many of these puzzles. In this informative and entertaining work, Michael Mott, whom I have had the pleasure of both knowing and collaborating with for many years, presents a beguiling and seductive range of anecdotes, accounts and traditions culled from the folktales and traditions of various countries and uses these to present persuasive evidence to support his hypothesis. When one thinks about it, it is perhaps the subterranean regions of the earth that are the only logical place these beings might actually inhabit, for almost all encounters with these entities feature their sudden disappearance into a cave or crevasses or even the ground itself. Certainly, there have been repeated accounts of a hollow Earth concealing fleets of UFO's and it also allegedly the site of Ultima Thule, the fabled source of the Aryan race, a place where man could, by whatever means leave the surface of the earth and journey deep into the bowels of the planet to an complex inner world illuminated by its own miniature sun.

My own inclinations are that this anecdotal evidence may have some support in hard science based on the existence of the 'Mohorovicic Discontinuity,' better and more conveniently known as the 'Moho'. This subterranean anomaly was first discovered by the Croatian geologist, Andrija Mohorovicic, and is a series of interlinked, city sized voids and cavities located approximately 30 kilometres below the surface of the planet, less below the ocean floor and the North and South

Poles. The Moho stretches right around the world and it may be the primary source of myriad myths and legends including the biblical Hell. Before dismissing this possibility out of hand, it is as well to recall the range of strange creatures currently surviving in much more inhospitable locations near volcanic outlets on the ocean bed and in other places that at first sight may appear incapable of sustaining life. When one combines the genuine existence of such an area and the possibility that life may have adapted to cope with the extreme conditions there and then reads of the legends and traditions collected and presented by Mike Mott, it quickly becomes clear that there may indeed be a viable ecosystem in existence right below our feet. I have no hesitation in heartily recommending this book to both the casual reader and the serious researcher.

Brian J Allan,

Author of **'The View from the Abyss', ' The Barriers of Belief' and 'Rosslyn, Between two Worlds'**

Fife,
Scotland,
January '07

Available from Hiddenmysteries.com

Foreword
by Dean Dominic De Lucia

I perceive that Caverns, Cauldrons and Concealed Creatures fulfills an important need, perhaps an urgent one. A world civilization which is hardly conversant with the true configuration of the planet on which it sits, with the creatures that live within it, and with the creatures and vehicles that fly in its skies, is certainly in urgent need of illumination.

We are under the impression that there is nothing left to discover. We could muse and say that Christopher Columbus would have been unemployed by the mid 1800s, if not earlier. By the end of the 21st Century we had even closely surveyed the surfaces of other planets in addition to our own. We have felt that there are no more worlds to discover, no more creatures to know.

But society has certainly received reports and information on cryptic creatures from all over the globe. There is no reason to doubt- there has been no lack of close up observations. There is even the record of an autopsy by a medical doctor of an Abominable Snowman from Central Asia who was shot in a cavern system in Afghanistan. The doctor was accompanied by a military officer and an entire regiment! So much testimony of UFOs has been received that there can remain little doubt as to their existence.

The obvious question is, where do they come from? It is doubtful that Chupacabras or Bigfoot live anywhere on the surface of the Earth, or else their habitats would have been found. And why would small sized UFO craft come all the way from the Zeta Piscium star system just to mutilate and drink the blood of a few head of cattle? And wasn't the Abominable Snowman killed in a cavern, wherefrom the other members of its clan escaped within? Caverns, Cauldrons and Concealed Creatures provides new parameters for our world because it reveals a vertical dimension below.

Granted that this world below has been ignored by mainstream institutions such as the media, governments, and scientific and academic institutions. But the information contained in this unique book would give a mid-course correction to entire fields of alternative study, such as the UFO community. UFOs have been photographed entering into the calderas of volcanoes in Southern Mexico. How far down did they go? To worlds below? Wouldn't these be a more likely origin for such craft than distant star systems? Bugs and insects inhabit the dirt and soil, and come crawling about when these are disturbed by plowing. Do the insect-like features and dexterous efficiency of the UFO occupants known as the Grays suggest that their origin might also lie in the depths below the surface of our planet? What salubrious conditions might exist in the cavern worlds below which would account for their size and advanced cerebral capacities? And why would such creatures come to the surface of our planet walking by foot? Of course they would have advanced vehicles to carry them back and forth.

But the UFO community is hardly aware of the issue of underground origins. Its leaders do not call the attention to underground origins but, rather, point their fingers towards the stars and deep space. Those that study Bigfoot and Sasquatch do not clamor for the probing of possible underground abodes- they seem to content themselves with horizontal investigation. Forteans may make note of the fact that ghosts only come out at night in the absence of the Sun, but don't make much connection to the caves and cavern systems below. There is certainly a crying need for this amazing book, which brings up the most obvious issue that others ignore.

Hollow Earth books argue a very nice case for the existence of a hollow cavity whose orifice is located in the high Arctic, and supportive evidence from the log books of the Arctic explorers of yesteryear is always presented. However, there is imbalance in this because the worlds within the crust and shell of the Earth are usually touched upon but briefly. Caverns, Cauldrons and Concealed Creatures completes the genre's presentation by offering a cohesive definition of the cavern worlds of the Earth's shell and its cryptic inhabitants.

This was lacking. Before the advent of Mike Mott's book, this type of information and understanding had to be pieced together. A true presentation the nature of our planet's outer shell and a variety of creatures now exists in an organized way, with supporting evidence and observations, in the form of this book.

In Caverns, Cauldrons and Concealed Creatures, though, the Cavern Worlds Theory has been born. This is a term coined by the book's author. Such a lofty achievement may be attributed to him because of the all-encompassing scope of his presentation and the strength of his revelations; such is the book's merit.

This book calls upon traditions from around the world to substantiate the existence of cavern worlds and their cryptic inhabitants. A more accurate definition of the wee folk is given which is faithful to tradition and folklore; not to the bedtime story books that we read our children. Fairies and elves are known for capturing healthy, young humans, for sucking out their life airs and leaving them old and unrecognizable, if they are left at all. They especially kidnap young girls and wives. Goblins were known for kidnapping and enslaving humans from the surface, especially for the purpose of mining activities and metalworking, deep below the roots of mountains. This book confronts the modern citizen of the world with the existence of hellish creatures that do pop up from time to time, something now lost to collective memory.

Though Christianity has dismissed its own tradition which tells of subterranean demons and devils, Mike Mott's book has established them again. Who else are the Grays, Chupacabras, and the other horrific-looking beings that Mike Mott offers evidence of? As long as we think of them as originating from distant star galaxies, we don't associate them with the demons told of by various scriptures, i.e., the Bible, the Koran and the Vedic literature. The Christian tradition does not describe demons riding high technology vehicles such as UFOs, but it can borrow from other traditions that do, such as the Hindu Puranas, which tell of the amazing craft of the demon Salva. In this way, the book confronts the reader with the fact that the UFO occupants are not affable space brothers but, rather, the demons which all good literature and traditions warn against.

In the pages that follow, the reader will even find the saga of Beowulf and Camelot reinterpreted according to the understanding of the Cavern Worlds Theory.

All of which leads to a new interpretation of the politics and economic practices of the surface world, referred to by the author as the underground manipulation of the surface world, which is yet another unique facet of this book. Sometimes it is hard for the mainstream individual to accept the idea of a worldwide conspiracy such as the New World Order is depicted to be. But if the surface world were under such sophisticated manipulation of the likes that only telepathic beings flying out of the inner world in flying saucers could mount, this would lend justification to the claim of a conspiracy by a mysterious, hidden elite. It would then be implicit that the international banking and business elite are not the ultimate cause, that there is a more remote cause behind them. In this way, the Cavern Worlds Theory shines a new light on the surface world's development of the last couple of centuries, and this is encompassed in the book.

Caverns, Cauldrons and Concealed Creatures

This is a book with a worthy message, worthy enough to cause a revolution in the thinking of modern minds. It redefines our understanding of the world around us, and schools us on our understanding of the cavern worlds below!

Dean Dominic De Lucia

Author of:

Astro Vedica
The Beatles & Vedic Astrology
Hollow Earth in the Puranas

Available from Hiddenmysteries.com

Foreward
by Jan Lamprecht

What would you rate as the worst disaster to strike Western civilization? The Black Death perhaps?

I have always thought it was the burning of the ancient library at Alexandria. A mentor of mine, Dr. Mitchell Gohnert, also felt that this was a terrible tragedy. We will never know the wondrous texts dating from Ancient times which were lost to our civilization forever and which perhaps stunted the growth of this civilization.

In my book, "Hollow Planets" (ISBN: 0-620-21963-7) I quoted Plato who spoke of a "God" who lives at the centre of the Earth. Would other texts in that library have told us more about this "God?" Would other texts have told us whether there are beings who live inside the Earth? Could these texts have led us to such a place?

Dr. Gohnert and I spent many an evening discussing the whole problem of the Hollow Earth at length. Was the Earth really hollow? Could it still be reached? We threw lots of ideas around. He suggested to me that no scientific study of the Hollow Earth would be complete without a reference to legends and to ancient literature. I dedicated one chapter in my book to this subject. However, I believed that several books could be written on this subject alone. This book by Mike Mott is therefore very welcome. I hope indeed that this book will be followed by others since the subject of scriptures and legends is very important. These scriptures and legends might yet lead to the discovery of subterranean life inside our own planet.

I have suggested, for example, that the Loch Ness monster might be a species of Plesiosaur which lives in underground caverns and lakes and which from time to time enters the Loch Ness.

The ancients may have known more about the interior of the Earth than we give them credit for. They may have learned some of this from more advanced beings (Gods?) they came across. For example, perhaps the Christian ideas of hell relate to lava and heat in the crust of a Hollow Earth? Perhaps the Tibetan Buddhist ideas of an Underground civilization relate to the "God" whom Plato said lived at the centre of the Earth.

Perhaps there are multiple advanced civilizations living underground in this Solar System who possess spacecraft which travel faster than light and which are virtually undetectable by our primitive science.

Since the publication of my book, I have been curious about the discovery by an American scientist, Dr. Louis A. Frank, that house-sized Icy Comets are striking the Earth's atmosphere in the Arctic several times a day. This does not make sense. Why don't these comets strike at lower latitudes since the Earth is racing along at 60,000 mph through space? Why are they limited only to the Polar regions? Then, when Dr Frank asked the US Navy to confirm his results they came up with negative results. Later, Dr. Frank learnt that the US Navy had deliberately mis-calibrated their instruments for the experiment. Why?

I have often wondered if the U.S. Military knows the location and entrance to a vast Underground Empire which exists here on Earth. I wonder if Dr. Frank (unwittingly) discovered the flight paths of space craft travelling into and out of a Hollow Earth by way of an entrance in the North Pole?

My own ongoing research into Arctic mysteries indicates that something extremely strange is going on up there in the Arctic sea directly north of Alaska. In this regard I have been invited to write a paper for the Society for Historical Discoveries in order to present the idea than an "unknown" island exists off the coasts of Canada and Alaska. Who lives up there? Eskimo rumours have it that Vikings from Greenland moved there centuries ago. Does that land exist on the rim of a Polar Hole?

In the study of legends and scriptures one must of course be careful who to believe and who not to believe. But as a general principle I do believe that there is no smoke without a fire. Much of what Mike Mott has discovered in his exhaustive studies has never been put together in this form before. I wish him luck in this undertaking, and hope he will continue with his research. This is an area which has been largely neglected and which needs some good reference works.

Jan Lamprecht
Johannesburg, South Africa
September, 2000.

Jan Lamprecht's Book is available at:

www.hollowplanets.com &

www.hiddenmysteries.com

THROUGH A GLASS, DARKLY

Introduction to the Second Edition

This book is dedicated to the idea that the structure of our planet, the history of our species, as well as the underlying truths disguised as our myths, legends, folklore and "pop culture" imagery, are not quite what we have been led to believe.

The Earth itself may be much more stratified and complex in structure than we have realized (or wanted to believe), and our species has perhaps experienced, through the ages, encounters and conflicts with inhabitants of other aspects of our planet's biosphere and electromagnetic matrix: "others," non-human intelligences or beings of various types who have seldom if ever been our friends. Memories and accounts of these encounters have come down to us in the form of folklore and mythology, often viewed with superstitious terror or awe through the conditioned lenses of religions, or religious interpretation.

For example, many Gaelic folk and fairy beliefs represent diminished forms of pre-Christian deities and heroes. If one studies the work of folklorists such as Evans-Wentz, Katherine Briggs, J.W. Campbell, and a host of dedicated others, it is found that, separate from the highly romanticized, expurgated, and sanitized Victorian versions of the same stories and traditions, the reality presents a much darker face indeed. The fairy-folk—even the lofty and lordly Tuatha de Danaan—were renowned as tricksters, deceivers, thieves, abductors and so on. The various folklorists who met with the common folk of the 19th century took their stories first-hand, got the exact meanings and interpretations from the common-folk (and sometimes, from the real percipient of the event), and this left no doubt as to this aspect of fairy-lore, even among the so-called "trooping" or regal fairies like the de Danaan. In fact, euphemisms were used for them out of obsequious fear of their wrath; hence they were called "the good neighbors," "the good folk," and so on. This kowtowing was to appease beings whose malice was greatly feared and whose mercurial wrath was easily aroused.

Caverns, Cauldrons and Concealed Creatures

As customary among the Highland Scots (see the work of J.W. Campbell, Popular Tales of the West Highlands, and The Secret Commonwealth by the Rev. Robert Kirk), great lengths were gone to in order to appease or placate the fairies and related beings. Food and milk offerings were often left for them, "their" places were avoided out of fear (called "respect"), and doors and windows were securely locked or latched at night. Some romantic folklorists wrote fanciful, dreamy verses about how wonderful, enlightened, and special these beings were, often creating purely idealized fabrications with little basis in the reality of folk-belief, as held by the common people of the countryside at the time. Often a great deal of sanitizing, along with sheer romantic fabrication, was the result. On the other hand, the Irish Book of Conquests, particularly the Early Mythological Cycle (the other sections are the Ulster Cycle, Fenian Cycle, and Historical Cycle), seem to recount the interactions of real and ancient beings and human beings, who warred over ownership of Ireland in antiquity.

According to ancient accounts from around the world, different types of beings were eventually driven beneath the ground where they dwindled in stature, to venture forth to prey upon genetics from the surface—human abduction, interbreeding and even marriages, and livestock and vegetable produce were all targets of their incursions. This much is wholly clear from the serious study of not just the Book of Conquests, but from the folklore of the common men and women of Ireland, which fortunately we have, thanks to the many days and nights of tireless efforts of many dedicated folklorists. This theme of "diminished giants," strange and mercurial "little people" and other opportunistic beings who dwell underground, avoid the sun, and live to some extent off of the prosperity of the surface-world ecosystem, are not just native to the Emerald Isle but are found worldwide.

The recent find of the so-called "hobbit" of Flores Island (given it's apparently carnivorous habits, perhaps "munchkin" would have been a more appropriate nickname) was not entirely unforeseen. In fact, such creatures were postulated in the first edition of this book.

As the recent find of Homo Floresiensis (the cave-dwelling, cannibalistic "hobbit" of the Indonesian Island of Flores) demonstrates beyond any doubt, hidden humanoids with both subterranean habitats, and nasty habits ("ebu gogo," as the islanders called the little people, abducted human infants in order to eat them, according the islanders), may belong more to the realm of a stealthy and hidden reality beside our own, than to just the areas of folklore and mythology. Folklore and fact overlap, and are one, in this instance. Perhaps this indicates that humanity has never, ever, truly been "alone". Could the descendants or relatives of the Flores creatures, both advanced and primitive, still live alongside us on this planet in hidden and even parasitic fashion today?

The verified 80,000 years of continual habitation at the Flores cave site should indicate that enough time passed during the existence of this species for it to have spread to other parts of the world, continuing to evolve, slowly but surely. According to natives of Flores, the creature still existed in significant numbers as recently as 300 years ago, when the islanders finally killed the cavern-dwellers in retaliation for kidnapping an infant; and other islands in the region have tales of similar beings, known as "orang pendek," a sort of midget bigfoot who steals crops. Some researchers are searching for living specimens of orang pendek even now, throughout the Indonesian island chain.

Additionally, brain-casts of the interior of the skull of the Flores creature, theoretically an offshoot of Homo Erectus,revealed that it had a primate brain unlike any other, with super-dense convolutions of the frontal lobe and surrounding areas, a sign of very high mental development. What abilities might this lead to over centuries or millennia? Telekinesis? Telepathy? Hypnotic abilities or "glamour"? In other words, the abilities of the "fairies," the "gray aliens," and other anomalous beings?

Upon inspection it is found that not only is the subterranean humanoids-in-hiding a universal theme, common in nearly or possibly every folk tradition, but it is echoed in the various mythological cycles and in the "great religions" of the world, stretching back into the most distant antiquity. Romanticism of terrors is just another way the human mind seeks to dismiss or alter—or sanitize—unpleasant facts for mental consumption. The Victorian take on "fairies" and the supernatural is highly suspect, yet has unfortunately tainted much of what most people think or know about this subject.

Now, instead of "gray neighbors," "trolls," "little people," and mysterious strangers like the Pied Piper child-abductor and the Plague-Bringer, we have "grey aliens," "reptoids," and "Men in Black." These are simply updated masks for the same ancient, parasitical "others" who have always been interested in human genetic wealth, and the prosperity of the surface "garden" that we know as our world. According to the serious literature (non-fiction) on this subject, "saucer people" run the gamut from seemingly-human to grays, hairy humanoids of various sizes and nasty dispositions, extremely unfriendly and violent "hairy dwarfs," reptilian humanoids, and even stranger forms like the "Flatwoods Monster" and the "Hopkinsville" or "Kelly Goblins." Yet here we see all of the inhabitants of the "secret places" of our world from the collective folklore of the planet, updated in presentation for the technological age. Throughout human history the inexplicable and frightening mysteries which confront us have been trivialized so that we can deal with the data and still function in our smug, human-centered, man-at-the-top of creation worldview.

From trivializing, ridicule follows, and thus the "terrifying" beings of yore are in the present day demoted to selling breakfast cereal with yellow stars and green clovers, or sliding down chimneys on Christmas Eve. No wonder they've changed their appearance or presentation to "pseudo-technological"! Now "gray aliens" and "bigfoot" are marketing caricatures, but this does not diminish the genuine terror, wonder, or amazement of real people who are certain that they have actually encountered or witnessed such beings. My theories and musings in this matter are based on two types of evidence: circumstantial evidence based on folklore, mythology, religion, legends, archeology, and related areas, and first-hand or witness accounts, both ancient and modern.

Often these two areas of source material or even anecdotal evidence overlap. I believe that these various phenomena and beings are in fact modern representations of those who have traditionally been considered fairies, gnomes, trolls, djinn, and so on, and their "magic" is a technology of such antiquity, advancement and complexity that it is, to our understanding, "magical." With UFOlogy, cryptozoology, fortean phenomena, and all the rest, we are in fact seeing, perhaps even participating in, an ongoing contemporary update of "folklore," which is in actuality the way our human minds attempt to gather, categorize, assimilate, and comprehend/decipher (as well as "cover over" with simplifications which mask our terror) the very real evidence for the existence of such beings and their ongoing yet seemingly inexplicable activities. □□In other words, it is all folklore, and ancient folk traditions may contain much more truth than has previously been believed, just as "forteana" and related topics do today! Once we make the leap of understanding that the creation of folklore is part of the process of comprehending these mysteries, and even may be caused by them, the whole picture of our culture, literature, fairytales, Jungian archetypes, and so on takes on an entirely different light.

The creation of folklore, based on very real "others" and their parasitical or otherwise peripheral activities, has become an instinctive, mind-protecting function for the collective human mind. I may not be the first person to make these connections, but I think that some of these concepts are original ones that need to be taken further, such as the idea of the creation of folklore, fairytales, and so on, being a built-in survival mechanism for the human psyche, which we have come to generate in order to protect our minds from things which are simply too horrific to contemplate or accept.

Therefore dismissal and ridicule are built into the system. This is, of course, not to our advantage. As the Apostle Paul remarked, we do indeed perceive hints of a greater reality, only seeing the shadows and flickerings of the mostly-unseen others which have always been with us; glimpsing them, and related mysteries, as "through a glass darkly." Smoky, hazy, and indistinct as this might be, some vision and understanding is preferable to none.

Wm. Michael Mott January, 2007

THE LONG NIGHT of HISTORY

or
Time for the Dreamers to Awake

By Wm. Michael Mott

Human history and culture are permeated and surrounded by long-standing traditions, folk-beliefs, and purported experiences of encounters, interactions, and conflicts with "others." These others are not human beings, as we would use the term, and do not always conform to those controlling conditions which we tend to think of as "natural law," or as more recently defined, "the laws of physics."

These "beings," seemingly existing at the periphery of our awareness or our reality, have gone by a variety of names: djinn, ifrits, fairies, elves, trolls, goblins, vampires, incubi....and more recently, as "little gray aliens," "reptoids" or reptilian humanoids, hairy humanoids, "nordics," and so on. While some seem to have a more substantial basis in physical reality than do others, a close examination of folklore, mythical, and historical accounts reveal that not only are all such forms inter-related in some way, but they have their origin in an aspect of reality which has been in large part ignored or dismissed as fantasy--a realm which nevertheless has a powerful presence, as far as the mythical and religious traditions of the world are concerned.

Throughout our oral and written history as a species, folktales and folk traditions have either hinted at or indicated the reality of this "hidden world" which surrounds us. The furtive and parasitical nature of this "alternate reality" leaves no doubt as to the intrinsic hostility of these "others," or at best, the indifference that some of their "racial types" or subspecies exhibit toward humankind.

From child-snatching tales (or distorted actual occurrences?) such as changeling legends, stories like those of Rumpelstiltskin and the Pied Piper of Hamelin, to the sexual assaults or predatory activities of succubi, incubi, and other manifestations of "demon lovers," and then onward to modern-day UFO abduction and hybridization accounts, the interest that these beings have--the NEED that they demonstrate--for fresh human genetic material or influx, becomes quite apparent upon careful examination. "Frog princes" and "fairy princesses" may in fact have their parallels in today's reptoids and nordics.

The reported fascination of UFOnauts with reproduction and interbreeding--as in the Villas Boas case--is surely pertinent to this topic. Are human beings--worshippers, servants, or unwitting fools--utilized upon occasion to assist in the procurement of genetic materials--people and livestock--in exchange for some sort of reward? The circumstantial evidence would seem to indicate that this also is a very real possibility. The reported "fairy" interest in human livestock—particularly cattle--and the extant tales from the British Isles and elsewhere of both cattle pilfering and genetic manipulation, and even dismemberment or exsanguination, are wholly identical to many aspects of the modern "cattle mutilation" phenomenon.

The visitations of the ancient Nephilim or Nefilim mentioned in Apocryphal lore, in which they presented themselves to human women for purposes of procreation or lust, are not at all dissimilar to the sexual depredations of fairies and modern day "aliens" of a wide variety of descriptions. Pattern and intent remain consistently the same, whether dealing with ancient "folkloric" forms or archetypes, or more recent ones.

Another interesting parallel is the tradition that the Nephilim and their partially-human offspring and descendants are imprisoned beneath the Earth, kept captive there, perhaps, by a changed surface-world environment and a low tolerance for the ultraviolet radiation in sunlight. Only those with more of the "stolen" human genetic material can venture forth for any length of time, which may in fact explain the seemingly desperate and unrelenting need for interbreeding, whether accomplished by trickery, seduction, or coercion. The same can be said for the "wee folk" of varying descriptions and traditions, as well as for vampires, evil spirits of various types, and a number of so-called "extraterrestrials" who have surprisingly little tolerance, for a "space-faring" race, for bright daylight conditions, even down to the dark protective lenses over their bulbous eyes. Great lengths seem to be gone to, in order to convince humanity that "they," the "others," come from somewhere else, when in fact their origin may be much nearer at hand than we realize. As for most humanoid creatures of "supernatural," paranormal, or cryptid aspect, sightings tend to be take place more often at night, or at twilight, dusk, or early morning, and in deep, dark locales such as heavy forests, misty valleys, and so forth.

In his highly-confabulated accounts of subterranean dwellers and their predations upon the surface world, the pulp writer Richard Shaver insisted that such beings had taken refuge beneath the Earth's surface as a result of a change in the sun, or in its effect upon the genetic structure of living tissue. We now know that prolonged exposure to direct sunlight and other types of radiation do in fact degrade our cells, and in fact may be the primary cause of aging. The Sumerian and Hebrew accounts of a flood seem to indicate that the atmosphere surrounding the Earth was once very different than at present, with an ever-present "canopy" or "firmament" which may have kept the majority of such radiation away from life on the planet. In fact, the Book of Genesis indicates that the "first rainbow" occurred after the downpour of the deluge (perhaps when the cloud canopy fell, due to some form of catastrophism), when the sun shone brightly upon the Earth. If this occurred upon a day which was, in actuality, the first day in millions of years without a hazy shroud of water vapor or even suspended ice crystals filling the upper atmosphere, then this may not be as much of a myth as many might wish to believe.

A rainbow would have been something new, and might have heralded the rise of a hardier species--homo sapiens--to a place of dominance in the surface world environment, and the retreat of older races or species beneath a protective ground. According to the records of Sumerian Kings and Biblical Patriarchs, the result of this exposure to increased cosmic and solar radiation had an effect on humanity as well, with life-spans dwindling, becoming shorter and shorter as years and centuries passed and our cells became less and less resistant to the bombardment from above.

The evidence for these facts seems at times overwhelming. Folklore, mythology, religion, literature, witness accounts--all seem to support this theory with a mountainous supply of circumstantial evidence. As we re-examine our own emerging global society, as we approach a century of great expectation for human development, we can nevertheless sense a tightening of "control" or restraint, like a leash, about our collective throat. The grinning despotic and de-humanizing skull of global governance seems to be only the herald, or sunlit surface-world manifestation of this sense that WE are not really in control of our own destiny, and never have been; that through human puppets who rule from offices, palaces and assemblies, we are all dancing or cringing to the pull of unseen strings, and this has always been the case.

The book CAVERNS, CAULDRONS, AND CONCEALED CREATURES examines these topics and the evidence for these musings in greater depth, with a close examination of most of the folk and mythic-religious traditions of the world and the relationships which emerge, as well as many first-hand, eyewitness or historical accounts. We have perhaps been so conditioned to accept the strangeness of it all--to dismiss the unusual or unexplainable as a delusion, fantasy, or item of ridicule and derision (which is actually a device to mask our own fear and uncertainty)--that we are in fact our own "worst enemy." Ignorance may be bliss, but the time may also have come to reach a state of full and empowered wakefulness and awareness as to what just might be going on.

Part I:
THE DEEP DWELLERS
Subsurface Inhabitants in Folklore, Myth, and Literature

PREFACE

The legends, myths, and literature of mankind have always been filled with fanciful or terrifying accounts of underground lands and races, hidden from surface sight. Rumors of a largely-unseen reality, of cavern-worlds, hidden tunnel systems which criss-cross the globe, and the occasional accidental discovery of a larger, geode-structure within the Earth have migrated from the realms of folklore and early scientific speculation, and into literature--and perhaps back into folklore again.

When reading and studying the available fiction which touches upon the topic of a subterranean world, many similarities come to light, which is interesting insofar as the various writers were not necessarily familiar with one anothers' works. It is obvious that many of them drew upon folktales and mythology, as well as the latest scientific findings and theories of the day, drawing indeed upon a huge matrix of archetypes and forms with which to work. Religious traditions have also been a major influence on the development of fiction about subterranean worlds and inhabitants, and some brave souls have shared accounts of what they have believed to be their own encounters with the denizens that dwell within the Earth's crust. In this work all of these aspects of underworld studies, and more, will come under careful examination, but this is not so much an examination of the underworlds as it is of their inhabitants.

SECTION 1: MYTHICAL (ANCIENT RELIGIOUS) ACCOUNTS

1. The East.

One of the earliest examples of subterranean stories is to be found in the Gilgamesh cycle of stories, which some would say is not so much fictitious as it is a distorted account of actual ancient events. Gilgamesh was an actual king who ruled Uruk (also called Unug) about 2600 B.C., and he was supposedly of half-divine origin. Like other heroes of ancient mythic cycles who were demigods, or semi-divine, Gilgamesh longed for an immortality which he saw as his birthright, much the same as Heracles of the Greeks. In one tale from his cycle, he befriends a physically powerful, hairy, subhuman character named Enkidu, and teaches him the customs of humanity. In a later tale, on behalf of his friend and king, Enkidu agrees to venture into the underworld of ancient Mesopotamia, to search for someone who has the secret of immortality. Other Sumerian accounts leave little doubt that the KUR, or KI-GAL (the "Great Below") was a place of immense size and great terror. This realm was ruled by the goddess Ereshkigal and her consort Nergal, a warlike god who had entered her queendom with plans of conquest, only to be seduced by the enemy, who became his wife. The KI-GAL was said to be filled with a wide range of beings, including the spirits and undead, reanimated bodies of human beings, and also savage guardians called "scorpion men." Other residents were described as sexless and robotic "artificial" beings called GALATUR or Gala, who were used by the rulers of the underworld for missions of kidnapping human beings from the surface world, or for other errands. Also present in the underworld were the UTUKKU, "eagle-headed" reptilian humanoids, which are probably the original djinn and ifrits of the ancient middle east. The latter beings are usually depicted with wings, representing their ability to fly when dispatched on the errands of the rulers of the underworld. Another strange race is the PAZUZU, a canine-faced, humanoid monstrosity with reptilian scales and tail. All of these are motifs which are found to permeate nearly every ancient underworld tradition, in one form or another, and have also found their way into folklore and literature.

Asia, particularly the Indian subcontinent, Tibet, Nepal, China, and Japan, have very similar traditions about subterranean inhabitants. In India there is a strong belief in the reality of the Nagas, a race of serpent-people or lizard-men who make their homes in two major underground cities (or civilizations), Patala and Bhogavati. The latter is said to be under the Himalayas, and from there the Nagas wage war on other, human, subterraneans, from the subsurface kingdoms of Agharta and Shambala. To this day, Patala is believed by millions of Hindus to have an entrance in the Well of Sheshna, in Benares. According to herpetologist and author Sherman A. Minton, as stated in his book "*Venomous Reptiles*," this entrance is very real, with forty steps which descend into a circular depression, to terminate at a closed stone door which is covered in bas-relief cobras. In Tibet, there is a major mystical shrine also called "Patala," which is said by the people there to sit atop an ancient cavern and tunnel system, which reaches throughout the Asian continent and possibly beyond. The Nagas also have an affinity with water, and the entrances to their underground palaces are often said to be hidden at the bottom of wells, deep lakes, and rivers.

The Nagas are described as a very advanced race or species, with a highly-developed technology. They also harbor a disdain for human beings, whom they are said to abduct, torture, interbreed with, and even to eat. The interbreeding has supposedly led to a wide variety of forms, ranging from completely reptilian to nearly-human in appearance. Among their many devices are "death rays" and "vimana," or flying, disk-shaped aerial craft. These craft are described at length in many ancient Vedic texts, including the Bhagivad-gita and the Ramayana. The Naga race is related to another underworld race, the Hindu demons, or Rakshasas. They also possess, as individuals, "magical stones," or a "third eye" in the middle of their brows, known to many students of eastern mysticism today as a focal point for one of the higher chakras, or energy channel-points, of the human(oid) nervous system--the chakra associated with "inner visions," intuition, and other esoteric concepts.

In China, the Lung Wang (Dragon Kings) closely resemble the Nagas in many respects. The Lung are said to dwell either in the "celestial realm," that is, the stars and planets, or beneath the surface of the Earth. They, too, possess a "magical pearl" in their foreheads, a "mystical" or divine eye or source of power. Like the Nagas, some of the entrances to their palaces or kingdoms could be found beneath lakes and rivers, or behind waterfalls. Almost always, such entrances are well-hidden from the intrusive eyes--or feet--of mortal men and women.

One such entrance to the Chinese underworld was said to be in the "Eastern Mountain" of Taishan, near Qufu province. This entrance to the Chinese Hell was guarded by savage demons called Men Shen, often depicted as warriors wearing fierce, animalistic masks or faces.

There was also interaction between the Lords of Hell, as they were known, and the Dragon Kings. The four Hell-Kings, called Yan Luo or Yen Wang (possibly derived from the Hindu death-god, Yama), rule over a vast region which consists of eighteen levels or locales. In one tale, a Dragon King is robbed by an extremely clever and human-like "monkey," who is similar in many respects to the Hindu Hanuman (who in turn came into conflict with the Rakshasas of the underworld, in the Ramayana). Enkidu also comes to mind in this tale. The Dragon King calls on the help of the Yan Luo, who are in turn bested by "Monkey" as well.

The eighteen regions of the Chinese netherworld are beaurocratic, tedious systems or civilizations, an apparent mixture of Taoist and Buddhist ideas, with a strong influence from the latter.

Japanese hell is even more of an amalgam of different traditions, as it incorporates Chinese and Hindu Buddhist characteristics into an older, animistic Shinto system. Emma-Hoo (perhaps from Yama-Raja) is the king of Jigoku, an eight-leveled region of fire and ice. Jigoku is filled with "Oni," or demons, which have the heads of oxen or horses, but humanoid bodies. Japan also has other cavern-dwellers who come out upon occasion, such as the bird-headed, reptilian goblins called Tengu, who dwell in mountainous regions, mostly come out by night, and are experts in the use of the weapons which are in modern times associated with ninjitsu. According to legend, the tengu trained Minamoto Yoshitsune, a famed samurai swordsman of the late 12th century, when he was an exiled boy on the run from his enemies. The Tengu are almost identical in description to the Utukku of Sumerian myth. Other beings are the Kappa, a semi-aquatic and totally reptilian-looking humanoid dwarf, the "ugly girls of hell," and sundry other shape-shifting. entities which dwell underground, or under houses. The Fox People are among the latter, often taking human form, and delighting in abduction and seduction of mortals.

Transitioning back to the west of Sumer, the underworld of Ancient Egypt had many things in common with its Mesopotamian counterpart. Called the Tuat or Duat, it was ruled by the God of the Dead, Osiris, the counterpart of Sumer's Nergal. It is the servants of Osiris, however, that are of concern here. One was the jackal-headed god Khentimentiu, and also Anubis, both gods of arcane knowledge, embalming, and other sciences. The god of knowledge, Thoth, was also a regular in the Tuat, and he had a humanoid form but the head of a baboon (which is in many respects very canine in appearance).

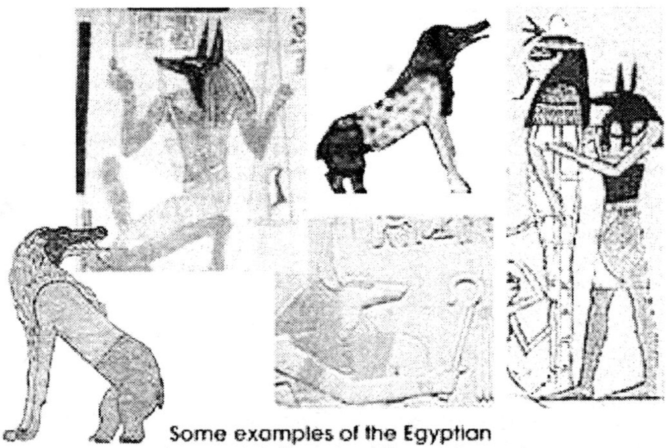

Some examples of the Egyptian underworld deities, Anubis and Ammut.

Some examples of the Egyptian underworld deities, Anubis and Ammut

(See Color Inserts)

All three of these deities bear a strong resemblance to the Sumerian Pazuzu. Another parallel exists in the Ushabtiu, originally conceived as artificial, animated and robotic servants which were very similar to the Gala, or GALATUR,, of the Sumerian underworld. Like the Gala, the Ushabtiu could be dispatched to punish or abduct an ordinary mortal, or even the Pharaoh himself. These beings were represented in burials by the Egyptians, with small statues of the same name, mimicking the supernatural servants of Osiris and hopefully providing a retinue of retainers and slaves for the deceased. Another being with familiar characteristics was the Ammut, a blend of crocodile, lion, and dog; but the most-feared being to haunt the underworld was possibly the god of chaos and strife, Set or Seth, who attempted to mount a coup de'tat against his brother Osiris. Set is often considered to be the Egyptian counterpart to the Judeo-Christian personality known as Satan, his appearance both canine and reptilian, with a human form but animalistic head, long-snouted or muzzled. The hugest dragon of the Tuat, however, was the gigantic serpent Apophis, very similar to Nidhoggr or Jormungand from the Norse underworlds, a monstrous serpent which brought fear even to the Gods.

2. The West.

Moving into Europe, the mythical subterraneans became less distant and more immediate, interacting with the common-folk on a much more regular basis. Scandinavians had their trolls/jotuns, also called etins (giants), which were great, granitic beings, sometimes hairy of form, and of immense physical power. More mysterious were the dwarves , a race (or races) of stunted, powerful craftsmen and weapons-smiths.

One race of dwarves was from Svartalfheim, the land of the Dark Elves, which was a cavern-world in its own right. Some of the Svartalf/Dwarf kind lived in Nifleheim as well, which was the land of the dead. Along with the savage and man-eating trolls, the dwarves would turn to stone, into toads, or otherwise die if struck by the direct rays of the sun. More often than not the entrances to their homes were hidden in inaccessible mountain sides and other remote locations.

The 'headless dwarfs' match the descripton of some UFO occupants, perhaps biomechanical robots. Abduction is the goal. Illustration: H.J. Ford

The Scandinavian and Germanic peoples also believed in the huldre or hidden folk, also called the Elves. Their domain was a luminous cavern-realm called Alfheim. From Alfheim they would venture forth to cajole, abduct, or seduce human beings. Other beings were the kobolds, or mine-dwarves, perhaps a variant of the Norse Svartalf. Another type was the Tusse, a variant of elf which lived close to humans, usually beneath a farmstead, or close to one. The primary interest of the huldre/elf-folk, which could be said to include all of the Germanic types, seems to have been procreation with human beings for purposes of maintaining genetic diversity. Like the trolls and dwarves, the elves seemed to dislike bright sunlight, but may have had more tolerance than their troll and dwarf cousins, as they were sometimes seen at dawn, twilight, or dusk, or by day in deeply-shadowed valleys or mountain chasms. Huldre/elves in particular were said to dwell beneath mounds and hills which were in closer proximity to human habitations, as trolls did more rarely.

The elves took a regular interest in human affairs-weddings, births, and deaths, (bloodlines?) the success of crops and livestock, and so forth—but only for their own selfish interests. They seemed to be overly-concerned with genetic and biological diversity, and they pilfered livestock, crops, and human genes via theft or cross-species liaison whenever they saw fit to do so. The elves are generally depicted as extremely fair-haired and fair-skinned.

Dragons were also said to live deep in the Earth, as recounted in the Anglo-Saxon (Germanic) saga of Beowulf; and the monster Grendel was a hairy, scaled fiend, a naga-troll which later British traditions would have called a goblin (and modern cryptozoologists would term a "hairy humanoid"). For Grendel, the term "Pazuzu" would probably have been just as appropriate. Dragons were the special guardians of "buried treasure," i.e., buried knowledge or technology, much of it often made by the powers of the subterranean dwarves. The European dragon had a nastier disposition than his oriental counterpart, or perhaps he came into conflict with a group of people who entertained different philosophical ideals when it came to living in fear of man-eating entities; whatever the case, like his eastern cousins, he had a relationship with both underground caverns, and deep bodies of water.

Caverns, Cauldrons and Concealed Creatures

England, Scotland, Wales and Ireland each have a rich tradition of under-earth dwellers, with many similarities or even common origins between them. Like the Norse/Germanic variants, the "fairies--" goblins, trows, knockers, brownies, leprechauns, sidhe (shee), tylwyth teg (terlooeth teig), and numerous other categories of humanoid beings--were fair or foul, malevolent or kind (actually, indifferent), making their homes almost without exception, beneath the ground. Mounds, hills, ruins, ancient raths or hill-forts, mountains, cliffs, and even cities of great age were said to serve as the rooftops of their palaces. Like the Nagas and dragons, some had the entrances to their subterranean homes at the bottom of lakes. To remove all doubt as to their relationship with Norse hidden-folk and Indian Nagas alike, they shunned the sunlight, and often seemed interested in crossbreeding their own bloodlines with those of human beings, or even in crossbreeding their "livestock" or fairy cattle, horses, hounds and so forth with the surface specie which were most compatible. The goblin-dwarf Rumplestiltskin, in his lust to have the human baby and it's genetic bounty, is just one example of this in folklore.

Of particular interest are the Tuatha de Danaan of Ireland, the "people of the Goddess Danu," also called the Sidhe. Originally an aristocratic, warrior race of heroic proportion, they dwindled in size after retreating underground, to become the Daoine Sidhe (theena shee) or diminutive "faeries" if Irish folklore. Most of the "gentry," or aristocratic, trooping faeries, are said to be of this type. Like the Scandinavian elves, they are depicted as particularly fair of complexion and hair. According to Lady Wilde, in "Ancient Legends of Ireland (Ward; Downey, London, 1887)," they are categorized as "cave fairies." In addition to their now-familiar practice of abduction of and hybridization with surface humans, their name, "Tuatha (tribe or people) de Danaan" holds a strong resemblance to the Egyptian name for the underworld, the Tuat. It is generally thought that the term "fairy" or "faerie" has it's origin in the earlier French term "fay," or the Latin "fatae," but the possibility of an older origin, as hinted at by the Tuat/Tuatha connection, may in fact bear further examination; for did not the "PHARAOHS" believe that they would journey through the TUAT on the way to their places in eternity?

The hills and glens of Ireland are also said to hide the remnants of at least three vanquished races: the Firbolgs, the Fomorians, and the Nemedians. All are ancient enemies of the Tuatha de Danaan, and were driven underground by the latter in the distant past, where they then dwindled in size (genetic diversity?) at an even earlier time than did their conquerors. The Fomors and the Firbolgs are probably the origin-race of many of the "bogeys" and pookas (bucas), goblins and hobgoblins, Scottish trows, and other malevolent, sometimes shapeshifting beings which seem to bear strong resemblance to the Scandinavian trolls, being perhaps a smaller variant. The etymological connection between "trows" and "trolls" is obvious, and reflects the sequence of both legendary and historical migrations to the British Isles, as well as the wars between each newcomer group with the currently ensconced one. Each group of faeries and goblins can be viewed, of course, as the dethroned, exiled gods of an earlier, defeated human culture. The question is, what were these "gods," which still exist in the popular imagination today? Were they symbolic pantheons or archetypes, or living beings which predated man on this planet?

3. The New World.

Native American cultures had similar beliefs in an extensive layered realm of caverns which was hidden beneath their feet. This murky world was believed to be inhabited by both human and humanoid beings, and by a variety of monsters and demons. Most tribes or nations had their own traditions of subterranean "little people," as well as other motifs, including reptilian or serpent-like humanoid beings. In addition to this, many tribes believed that they had themselves emerged from a mythical underworld, ages before.

The Mescalero Apache have many of these beliefs. One of their oldest sacred traditions states that they came from the "Old Red Fire Land," before the "Great Flood." This land was said to be in the distant eastern (Atlantic) sea, and was destroyed by a combination of deluge and volcanic cataclysms. Escaping through "great caverns" and tunnel-systems, the ancestors of the Apache came to high mountain lands far to the south, where they built new cities. A series of misfortunes there, however, eventually drove them northward.

This is nearly identical to the origin story as related in the Mayan Chilam Bilaam, and brings to mind both Mayan and Aztec origin myths. The Aztecs said that the had originated in a land called "Aztlan," obviously synonymous with Atlan or Atlantis, the destruction of which they also escaped. After this, they ended up in a cavern-world called "Chicomoztoc," or the Seven Cavern Cities of Gold," where they lived for some time before emerging again into the surface world.

In addition to believing in a vast, nine-layered underworld filled with strange beings, the chief god of the Maya, called Itzamna (meaning "iguana house"), was depicted as an anthropomorphic lizard, snake, cayman, or dragon. The underworld dwellers were a mixture of human, reptilian, and other animal characteristics, and the rain-god Chac was a long-nosed, fanged humanoid creature very similar to the Egyptian Set. The underworld, called Xibalba, is the location of most of the action in the Popol Vuh, a priestly epic of the Maya.

In the Popol Vuh, two semi-divine brothers, Hunapuh and Xbalanque, have to journey into a realm of horrors beneath the Earth in order to defeat those who are enemies of their father and his family, and to their own ascendance to power. In Xibalba, they come into conflict with Zipacna, a crocodile-headed monster, Seven Macaw (who is bird-headed) the maker of earthquakes, and other familiar forms. An interesting event occurs when, seeking vengeance by destroying the lords of the underworld, the brothers devise a way to pick out the twelve lords of Xibalba from identical "mannikins," or robot/ushabti-like figures. The twins eventually defeat their underworld rivals, and take over the rulership themselves, bringing an end to human sacrifice as part of the deal. These events preceded and made possible the "modern" epoch of time.

The Hopi of the desert southwest, descendants of the mysterious Anasazi people, have an equally strange tradition. They believe that as a people, they migrated from a series of previously-extant "worlds," usually interpreted as "ages" or "epochs;" but these are also seen as subsequent cavern-worlds, each one lower than the next, each one eventually abandoned and destroyed in turn. While still in the murky "third world," the Hopi ancestors came into contact with the mysterious "Ant People," an ectomorphic race or species which greatly resemble the gala, of Sumerian myth. At some point they also came into conflict with the "Serpent People," and like other tribes, their underworld mythos is filled with cataclysms and floods. At least one of their previous worlds was said to have been "in the east," and combined with the flood element, is very similar to Apache and Aztec traditions.

The Choctaw (Cha'ta) people of Mississippi also have a myth of underground origins. They believe that their ancestors emerged from the Nanih Waiya Cave Mound, a fifty-foot-tall natural geological formation which is hidden in a swampy forest area, approximately a mile and a half east of a better-known, artificial mound and tourist site. The hill has several natural openings, some of which have been "sealed up" (the Park Service seems to have no good explanation for this), and it is said by the Choctaw to be the entrance to a vast underground realm. One legend has it that, in ancient times, the Choctaw were invaded by a race of red- and blond-haired, white-skinned giants, who bore "sharp clubs," (swords?) and axes, and wore an extra, thick skin (chain or leather mail?) which made them impervious to arrow, spear, and warclub. Add the touch that some of these Nahullo, or giants, "had horns," and these white invaders sound suspiciously like wandering Norsemen.

Whatever their origin or identity might have been, these invaders drove the Choctaw into hiding, and the Indians went into the cave mound for several generations. The world beneath the mound was a large series of caverns, through which a river or rivers ran (the Nanih Waiya cave mound sits squarely at the headwaters of the Pearl River). Some traditions indicate that it went on to connect to other "worlds," or underground places. Staying underground for many generations, the Choctaw emerged to wage a form of guerilla warfare on their enemies, eventually winning, by using darts coated with a poison made from mushrooms found in the caverns. Victorious, they emerged again into the sunlit world.

One tradition holds that this emergence of a generation of people who had been born underground is the basis for the mound origin myth, and that in fact the Choctaw had arrived centuries earlier, after leaving a "sunken land" which had foundered in a distant western ocean. After many wanderings and travails, they arrived in the southeast, where they found the natural cavern mound which would later serve as a place of refuge. But other Choctaw beliefs dispute this, claiming that not just the Choctaw but the Muskogee, Cherokee, and Chickasaw peoples emerged from the mound as well, having all been one people in the underworld.

Today the Choctaw still believe that a variety of strange supernatural beings either inhabit the cave-mound, or dwell in the wooded hills that surround it. One of these is the Shampe, a hair-covered, manlike giant who has a terrible odor;, and who stays underground during the day. The Shampe is a sort of Sasquatch, but the underworld connection is there. Also present are the Kawana-kasha, (Kowi Anukasha), also called Bohpoli (stone-thrower), a type of supernatural and mercurial dwarf; these live within not just the woods of Mississippi, but within the cave-mound itself. Like the Norse dwarves, they are the hoarders of vast knowledge. The mound is also the home of "giant serpents," and perhaps a host of other beings. Among the latter is the Nalusa Falaya, or "Long Black Being," who is humanoid yet slides on his stomach "like a snake." His pointed ears only accentuate his reptilian appearance. Another variant is the Nalusa Chito, "Big Black Being," who emerges from underground dens to capture women and children, presumably for supper. But this abduction scenario is by now a familiar one, and is very similar to the abduction and changeling accounts of Celtic and Scandinavian traditions, which were often for purposes of maintaining genetic diversity. The goblin "Ho'koklonote' she" is a shapeshifting creature believed to haunt the region, and is very similar to the Pooka or Buca of the British Isles. So are the "Nalusa" twins, for that matter.

As has been demonstrated, many similarities, or perhaps identical descriptions, exist for the underworld inhabitants of myth and folklore. This underlying cohesion may have resulted from an "archetypal stew" which long-simmered in the imaginations of men and women; and as will be demonstrated in the next section, this has resulted in some very interesting and imaginative works of fiction.

SECTION TWO: LITERARY EXCURSIONS AND REVELATIONS

Part One: European

FOUNDATIONAL FORMS

1. Wagner.

In Richard Wagner's operatic re-telling of the Volsunga Saga, he takes the audience on a tour of the German version of the Norse underworld. In the Nibelunglied, he tells the story of the theft of the Gold of the Nibelungs, a race of Dwarves from Nibelheim, which is the Norse Nifleheim, complete with it's hidden knowledge/technology and imprisoned Powers of death and darkness. The dwarf Alberich (a variant of the older Norse/Icelandic Andvari) and his kin are miserable underlings and errand-boys for deeper, more mysterious beings. They are the same type of transitional beings which the underworld gods of other cultures sent to the surface world on dark errands, and as such remind us of the Sumerian and Egyptian "artificial" (genetically-engineered) forms, and of the same beings as created by the Hebrew "Nephilim," or "Fallen Ones," which were rebellious angels/titans/jotun-trolls. The similarity between the Hebrew plural noun Nephilim and the Norse Nifleheim is perhaps more than coincidental. Knowledge and power in the Wagnerian opera take the form of a magically wonderous "ring" of dwarf or faerie gold, "Das Rheingold."

Additional obvious similarities include "giants (jotuns/trolls)," and semi-reptilian beings like Fafner (Norse Fafnir), who begins as a humanoid titan and is transformed fully over time into a cave-dwelling dragon. Wagner's opera-cycle is essentially a re-telling of the Volsunga Saga, with other elements borrowed from Norse traditions as well. The value of examining Wagner's work lies in the fact that it provides a bridge between ancient Nordic mythical traditions, and later literature which used similar motifs. These will be examined at length very soon.

(See Color Inserts)

2. Dante Alighieri.

Of equal interest is the earlier work of Dante Alighieri, in particular that part of his Divine Comedy called the Inferno (The Project Gutenberg Etext of Dante's Inferno, August 1997, Translanted by Henry Wadsworth Longfellow). In this tale, Dante is taken on a personal tour of the netherworld by the ghost of the Roman poet Virgil, and it is revealed that, like most mythic traditions, Hell is multi-layered, in this case with ten levels arranged in a conical fashion which diminishes in diameter as it descends. These concentricities are designated areas of imprisonment and punishment for specific types of sinners, and are the abodes of giants, harpies (another psuedo-bird/reptile form), horned demons, and so forth. The similarities between these forms, those of Greek Erebus and Tartarus, the Sumerian tradition, and even the of the Oni of Japanese Jigoku, are readily apparent. The deeper the descent, the colder the conditions become, so that sulphurous and hot, rainy, misted, and frozen conditions are all presented, as in the Norse, Greco-Roman, and Hebrew (Apocryphal) traditions. At the ultimate, frozen bottom of this happy land is Satan himself, frozen solid in a lake of ice, awaiting the day of judgement. This is reminiscent of the Norse and Greek cavern worlds, with regions and rivers of fire and ice near or adjacent to one another, where the Jotuns and Titans lay imprisoned in chains of darkness and entropy, only able to send out their underlings to work their will upon the Earth.

Illustration for Dante's Inferno, by Donn P. Crane

(See Color Insert Page 1)

3. Jules Verne.

Verne's Journey to the Center of the Earth, apparently based upon hollow-earth theories, rumors of underground tunnel systems, and the latest scientific thought of his time, was also filled with levels of increasing complexity and diversity, fire and ice, and was inhabited by the "latest twist" on ages-old underworld reptilian themes--the dinosaurs.

The lifestyle of the subterranean hybrids
Dwarves, elves or the slave-drones of another Race?
Illustration: H.J. Ford

These creatures, for Verne's purposes, replaced dragons, Nagas, trolls and goblins as the denizens of the depths. While this did not do justice to the rich heritage of traditions available to draw upon for such a work, Journey to the Center of the Earth, may well be the first truly modern novel to take on the subject of inner or hollow-earth mysteries, an inhabitants. It was the first to utilize both scientific principles and the latest scientific theories and discoveries as fundamental elements of the plot and storytelling. It still fell far short of the scope of later works on the subject, however.

FAERIE LANDS REVISITED

4. George MacDonald.

In the 1880s, George MacDonald published The Princess and the Goblin in Britain. Drawing mostly from "faerie" and "goblin" traditions of the British Isles, he spun a magical tale of a mysterious and secret underworld, inhabited by inbred, deformed dwarves called "goblins." Dwelling in a veritable maze of tunnels which seemingly underlay the countryside, these beings shunned the daylight like their Scandinavian and British Isles counterparts. They also shared their folkloric cousins' interest in abducting and interbreeding with surface humans, as in the case of Princess Irene, who attracted the attention of the King and Queen of Goblin Land as a suitable match for their "reptilian-" appearing son. This "goblin prince" was himself the result of a liaison between his underworld father, and a surface woman whom he had kidnapped, and who had eventually perished. Undertones of this sort run throughout the book, as in the Goblin-Queen's hiding of her shameful "deformity," a very human set of toes on each of her feet which spoke of human genetics somewhere in the mix. Like the kobolds of Germany and the knockers of Britain, the goblins were heard to tap and knock deep in the mines by human miners, when they would dig too close to their homes. Like the Nephilim, like the faeries, and like the underworld gods of the middle and far east, MacDonald's goblins had also bred an army of genetically-mixed, deformed animal creatures, which combined the traits of mammals and reptiles, much like the Pazuzu, or the Egyptian Ammut.

5. J.R.R. Tolkien.

Caverns, Cauldrons and Concealed Creatures

Another British author whose work contains "cavern-world" themes is J.R.R. Tolkien. Like MacDonald, he drew on Norse, Germanic, and Celtic folk and mythic traditions for much of his cosmology. His "orcs" or goblins, are essentially the "dark elves" and trolls of Scandinavian myth, and he also took his version of the dwarves from the same folk traditions. In Norse myth, there seem to be two camps of dwarves, one dark, swarthy, and hostile (like Andvardi / Alberich), the other reclusive, but not so nasty. Tolkien's "dwerrows" or dwarves fall into the latter category. His "elves" are similar to the Scandinavian variety, but actually more closely resemble "the gentry" of Ireland, the aristocratic Tuatha de Danaan. Like the Danaan, they also exhibit idealized "Aryan" or "Nordic" attributes. Deep mountain tunnels and hidden cavern palaces are the haunts of orcs, elves, and dwarves alike.

He also made reference to the "mound folk," in this case the undead (similar to Scandinavian "draugs" or revenants), and to "trolls," who, like their Norse counterparts, cannot stand the light of day. His hill-burrowing hobbits were modeled in some respects after both the 'trooping faeries" of the British Isles, and the "solitary faeries" such as leprechauns and cluricauns (see the in-depth works of Lady Gregory, Lady Wilde, and Katherine Briggs, on the "faerie" or fairy topic), and the brownies (who also have hairy feet like hobbits--and hairy hides in general).

His orcs in particular are generally reptilian in appearance, yet humanoid, and "Gollum," a hobbit who has been horribly mutated by the "radiation" of the One Ring, is definitely amphibian or reptilian in aspect. Here is an echo of Alberich again, in Gollum's lust for the ring, and here also is the hint of "technology," i.e., ancient and mysterious magic, which can genetically alter living beings for evil or unknown purposes. One side-effect of the ring is the extreme prolongation of Gollum's lifespan (another J.R.R. Tolkien "secret," that of immortality, as in the story of Gilgamesh and Enkidu). Throw in cavern-dwelling dragons guarding "treasure," (i.e., secret knowledge), and the archetypal symbolism is complete and in fact predictable. He uses all of the aforementioned underworld themes in his books The Hobbit, The Lord of the Rings Trilogy, and The Silmarillion (all published in the U.S. by Houghton Mifflin Co.).

Part Two: North American

The twentieth century brought a host of new writers, each mixing fantasy, science, and myth to some degree, all drawing on the wealth of lore and folktales which had come before. Most of those whose works reached a wide audience were American writers, working for the "pulp" (so-named for the low quality of paper they were printed on) magazines like Weird Tales, Amazing, Argosy and many others. After publication in these periodicals, a select few, like Edgar Rice Burroughs, went on to successful book publication of their serialized "pulp" features. The writers to be examined here were more or less divided into two camps: those who wrote about hidden cavern-worlds and tunnel-systems and their inhabitants, and those who dreamt of grander concavities in the form of hollow moons and planets.

CAVERNS, GROTTOES, & ELDRITCH HORRORS

1. A. Merritt.

In 1923, A. Merritt first published his excellent work The Face in the Abyss, which was reprinted in the October 1940 issue of Famous Fantastic Mysteries. In this fast-paced adventure tale, a group of American treasure-hunters search for a lost treasure in the Peruvian Andes, and find a "lost race" instead, the survivors of "Yu Atlanchi," an ancient Aryan"/Atlantean colony which has retreated into caverns and chasms beneath the Andes. In addition to the lost treasure of the Incan king Atahualpa, the hidden people have another secret or two: they are worshippers of a subterranean entity called the "Snake Mother," and are guardians of a bottomless cavern abyss which is overlooked by a cliff-face of solid gold, which has been formed into the likeness of a gigantic and Satanic face, reminiscent of the frozen Satan of Dante.

Add to all of this some remnant dinosaurs, flying, intelligent "serpents," and a half-snake, half-woman entity, and the age-old reptilian-underworld connection is complete.

2. H. P. Lovecraft.

Another well-known writer of pulp fiction who has not only stood the test of time, but has amassed huge legions of readers and fans, is Howard Phillips Lovecraft of Providence, Rhode Island. Writing mainly for Weird Tales, Lovecraft created a somber and terrifying universe of eldritch horrors, ancient and alien intelligences masquerading as demons or dark gods, most of them locked away or imprisoned beneath the earth or under the sea (like the Nephilim, Titans, Jotuns, and so forth). From their deep prisons they supposedly still haunt mankind, sending forth genetically-altered underlings and unhuman fiends they have created, to work their will upon the surface world. All of this is of course very, very familiar. There is a direct parallel here with the Hebrew's imprisoned Nephilim, and their partially-human or partially-animal, genetically-engineered offspring, as described in the Apocryphal books of Enoch, and of the Sumerian underworld rulers, with their robotic, reptilian, or otherwise genetically-altered humanoid slaves.

Lovecraft described a variety of underworld fiends and horrors, such as dog-headed humanoid ghouls from the depths of the Earth Pickman's Model, Arkham House, 1939, 1945), and eons-old invertebrate horrors lurking in a forgotten cavern world beneath Antarctica (At the Mountains of Madness, Arkham House, 1963), and many others. In The Call of Cthulhu, Lovecraft introduces the arch-fiend of his cosmology, a buried horror of humanoid yet squid-headed configuration, of titanic size, who sits buried in a sunken city in the depths of the Pacific ocean and directs all manners of horror by means of telepathy and messengers sent to the surface world. The parallels to Anubis, Thoth, the Ammut, the Pazuzu, the Greek Gorgon (an underworld creature), and so forth are apparent.

There has been some speculation that Lovecraft believed what he wrote to be true on some level, due to his descriptive dreams of "abductions" by beings he referred to as "night-gaunts," winged, reptilian humanoids without facial features who sound suspiciously similar to the Galatur or Gala, of the Sumerian KI-GAL. His fiction, he believed, was in some way influenced by "revelations" he received during these "dream-abductions," which is a very familiar story to those who study the modern

"UFO Abduction" phenomenon. But perhaps more prolific on the theme of subterranean races and horrors was Lovecraft's friend and contemporary, Robert E. Howard.

3. Robert E. Howard.

Robert Ervin Howard was born in Peaster, Texas, in 1906, and grew up to live in the oil-field town of Cross Plains. He only lived a scant thirty years, dying by his own hand in 1936. But during his thirty years he was one of the most prolific short-story and novella writers of his day, filling the pulps with characters which have long-since passed into popular culture as icons or stereotypes. His most-famous creation, Conan the Cimmerian, has been redefined and over-simplified for the world as Conan the Barbarian; but despite the distortions and over-simplifications of non-prose media (not to mention the input of other writers who missed the character entirely), the mighty Conan is still popular today. Howard is largely credited with the invention of the genre of "sword and sorcery" adventure fiction (even though Fritz Leiber coined the term), a claim which seems to be accurate. Although Conan met and dispatched his share of subterranean horrors, it was Howard's other fiction which seemed to concentrate on those themes so common to the underworld tradition.

Dog-headed humanoids, reptilian "toad" gods, unspeakable monstrosities from wells, pits, and endless catacombs, all of these and more flowed from his mind with a raw, emotive power. In Worms of the Earth (Weird Tales, Nov. 1932), is seen one of his favorite themes, which deals with a "lost race" which has lived for ages underground, mutating into something no longer human, yet longing for human genes, and the destruction of humanity. Other stories, like People of the Dark (Strange Tales, June 1932) and The Children of the Night (Weird Tales, April-May 1931), deal with the "racial memory" of ancient races of stunted subhumans who have retreated underground, to evolve into serpentine form.

(See Color Inserts)

In The Valley of the Worm (Weird Tales, Feb. 1934), the story is that of Niord, a wandering " Aryan" " tribesman of eons past, who comes face to face with an ancient, underworld horror which must be destroyed, in the form of a massive, sentient worm-like being from the depths of the Earth (a dragon or serpent god), and its hair-covered, faceless, humanoid servant, reminiscent of both satyr and sasquatch. Tellingly, the latter plays the pipes of a Pan or other cave-dwelling satyr.

Perhaps Howard's greatest "underworld" tale, however, came from his "King Kull" or "Kull of Atlantis" series of stories. In The Shadow Kingdom (Weird Tales, 1929), Kull, usurper-king of Valusia, becomes aware of the fact that most of his councilors and government officials have been "replaced" by imposters, which are revealed to be agents of "the-snake-that-speaks," or serpent-men. These are reptilian humanoids from a cavern/tunnel underworld, who have the power to assume the appearance, and the place, of any man or woman in the daylit, human world. They were identified by their inability to speak a specific series of syllables, a "code phrase" which human beings could use to weed them out and kill them. All of these themes go back to the most ancient underworld traditions, the "abductor-changeling-reptilian" servants of evil all rolled into one. This exact theme has also become prevalent in the conspiracy and UFO literature of the 1990s, where some would have us believe that not only is it true, but it's "newly revealed" or discovered; but the archetypal roots for the idea stretch back for millennia, and even pulp fiction has used it with great effectiveness.

HOLLOW MOONS, HOLLOW WORLDS

4. Edgar Rice Burroughs.

Perhaps the most well-known of American fiction writers to approach the Hollow Earth or underworld topic was Edgar Rice Burroughs (September 1, 1875 - March 19, 1950),. The creator of Tarzan of the Apes, Burroughs was highly-prolific, writing on a myriad of topics, and of a myriad of worlds. His straight-forward, slightly Victorian adventure-writing style made him the undisputed favorite writer of fantastic fiction in his day. His first novel was set on Mars, and from there he ranged from the jungles and lost cities of Africa, Asia, to Central America, and the back-alleys of Brooklyn, to the planet Venus and "Beyond the Farthest Star." However, his hollow earth endeavors began with "At the Earth's Core," written in 1914.

Burroughs begins by taking a pair of adventurers—Abner Perry, inventor, and David Innes, business partner, on a "maiden voyage" of the tunneling transport, "The Iron Mole." However, something goes wrong with the contraption, and after a near-death experience miles beneath the crust due to extremes of heat, they eventually find conditions in the machine returning to normal, and they break out into an unknown world. This is "Pellucidar," as they name it, the interior of the hollow earth. Obviously inspired by the non-fiction books of John Symmes, William Reed, and Marshall Gardner, Pellucidar is a land of eternal sunshine, with a small central sun ever at high noon. It has openings at both poles, and a lush, tropical terrain, with a land extent far beyond that of the outer world--for it is a world inversed, with landmasses which roughly correspond to our ocean areas, and vice-versa.

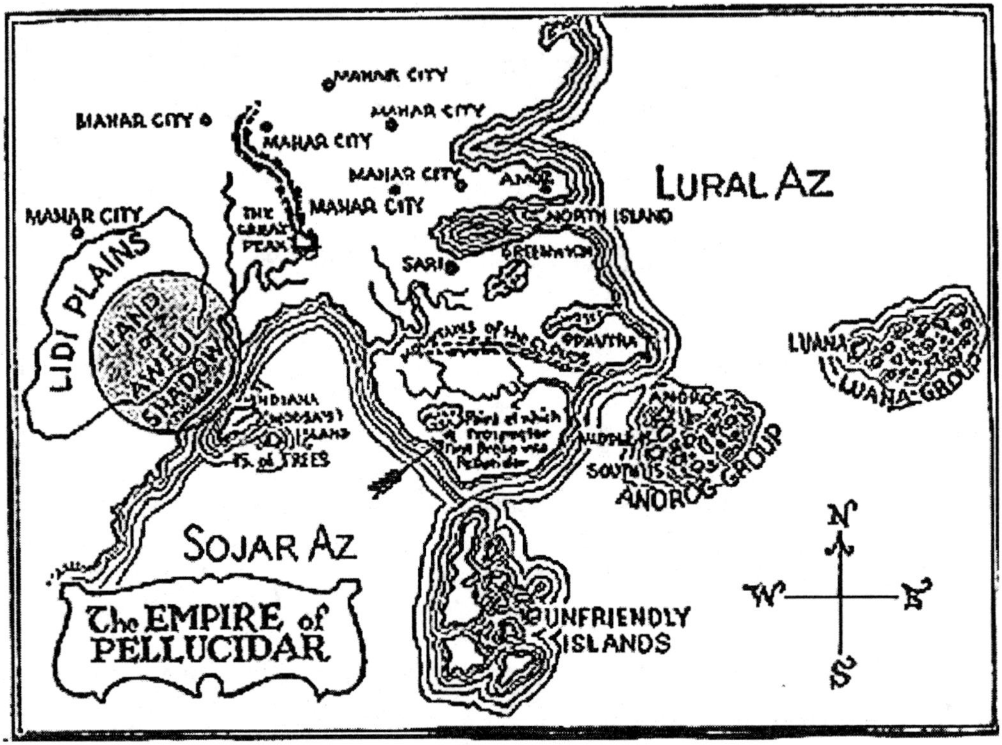

Other interesting touches to the series (in Tarzan at the Earth's Core, 1929) include an expedition by dirigible, through the north polar opening, probably inspired by the real-world journey of the dirigible ZR-1 in 1924 (Popular Science, December 1923), which was mounted by the U.S. Government in order to discover the myth or reality of the “polar continent” (or “opening?”). The results on the real expedition however, were never made public. There is also an interesting “pendant world,” a moon which revolves around the central sun in perfect synchronous motion with the rotation of the Earth itself, thus casting its shadow continually over one dismal region, the “Land of Awful Shadow” (Land of Terror, 1944). Another strange side-effect of Pellucidar is that it's inhabitants do not age, much like those of fairyland, or the Sumerian underworld of the gods. This is either due to the fooling of the "biological clock" of the body (it's always set at noon), or to protection from cosmic radiation which reaches the surface world.

Burroughs wrote five books (one of them a group of four novellas), mainly telling of the adventures of stalwart Americans and Europeans in a world which is not only in the stone age, but is in fact inhabited by dinosaurs, gigantic mammals, and strange unhuman races. Humans are indigenous as well, and some are soon taught the ways of the "wiser" surface men, and under Innes, form a benevolent empire. Before raising cries of "white colonial oppression!" however, the reader must understand the inspiration for forming allegiances and empires in the land of the eternal sun: There are enemies to vanquish, enemies which fragmented stone-age tribes cannot defeat unless they combine their forces and their determination.

These enemies include the Mahars, a race of "rhamphorhynchus," or intelligent pterandon bipeds, which feed on human flesh. These creatures, not only telepathic, are capable of mesmerizing nearly any human being into submission before having him or her for lunch.

They are all female, and reproduce by means of something called "The Great Secret," a form of cloning or other reproductive technology. Here we see another incarnation of the Nagas and the utukku, complete with an emphasis on reproduction. These reptilian overlords--er, ladies—have in their employ another group, a bunch of sub-human louts called Sagoths, hairy and barbaric, who raid human villages to supply victims for their scaly mistresses. Here of course is Enkidu again, or Sasquatch, or Grendel....

Other interesting races in Pellucidar include the Horibs, reptilian humanoids who also feast on humans, "little men," and the Gorbuses, who live in underground caves. Throw in a load of dinosaurs of every description, along with human-animal hybrids, and Burroughs has created an archetypal paradise of adventure and terror, fun for young and old alike.

A reptilian-human hybrid, or 'troll'
Illustration: H.J. Ford

The books in the series are:

1. *At the Earth's Core* (All-Story, 1914)
2. *Pellucidar* (All-Story Caval, 1915)
3. *Tanar of Pellucidar* (Blue Book, 1929)
4. *Tarzan at the Earth's Core* (Blue Book, 1929)
5. *Back to the Stone Age* (1937)
6. *Land of Terror* (1944)
7. *Savage Pellucidar* (Amazing Stories, 1963)

Burroughs also wrote three additional novellas along similar lines, The Moon Maid (Argosy Weekly, 1923), The Moon Men, (All-Story, 1925) and The Red Hawk (All-Story, 1925). These are set in the hollow interior of the moon, and are a fun read, but are somewhat more fanciful than the Pellucidar series. The Pellucidar books are amazing in that they are the first works of fiction to fully integrate modern ideas and theories about the Hollow Earth--i.e., the Earth as a geode-like structure--with the reptilian archetypes which have haunted the underworlds of the human mind throughout the centuries.

Other notable mentions of Hollow Moon stories are those of H.G. Wells (The First Men in the Moon, London, 1901, and McFarland & Co., 1998.) and Everil Worrell (The Hollow Moon, Weird Tales, May, 1939). Additional books of interest which touch on "amphibian (reptilian-galatur,) humanoid" and "Aryan" /Ascended Master" themes are Etidorpha (John Uri Lloyd, Dodd, Mead, & Co., 1901) and The Coming Race (Edward Bulwer Lytton, Routledge & Sons, 1871). Another book, The Smokey God (Willis George Emerson, Forbes & Co., 1908), purports to be a true account of a father and son who were sucked into the polar opening at the top of the Earth, and were the unwilling guests of a race of "Nordic" or "Aryan" "giants," who more closely resemble descriptions of the Norse Gods and their Jotun enemies than anything else.

PERSPECTIVES

So what do we have so far, if we combine folk traditions, mythological and religious traditions, and literary constructions? At first glance, it might seem disparate at best, but the inhabitants of the various underworlds remain surprisingly consistent in their general forms and dispositions, when examined more closely. We can start with a basic look at their physical and other attributes:

1. They are more or less reptilian, sometimes mostly human, sometimes very scaly indeed. Others, perhaps the result of liaison or gene-mixing with surface humans, are described as "fair," "nordic," or " Aryan"." The true origins or identity of the latter are generally betrayed by some inherent flaw of form or behavior.

2. They are telepathic, hypnotic, or possessed of other "superior" mental powers.
3. They can create "glamour" or illusion, either as faeries, or as shapeshifters.
4. They possess superior technologies, including cloning and genetic engineering, in some instances manufacturing artificial lifeforms to do their bidding. Some of their servants are often humanoid and hairy, others are scaled or even winged, most are bipedal and anthropomorphic. They possess the secret of immortality, or the formula for extreme longevity. Some possess the power of flight as well, via either aircraft or by self-propelled means.
5. They are interested in human reproduction and interbreeding, or in reproduction in general. Reproductive survival seems to be essential to their agenda in nearly any given scenario. They need human flesh, or blood, or reproductive materials.
6. They are more often than not the enemies of mankind.
7. They do not like the light of the exterior sun (Sol). As a general rule, they cannot survive for long, or sometimes at all, in the surface world.
8. They are banished, exiled, imprisoned, lost, or in hiding from the surface world, or from the sun.
9. They are secretive about their entrances and their treasures or knowledge, as well as their true nature or identity.
10. They influence human events or circumstances whenever it suits their purpose, or appeals to their whims.
11. They have surface humans in their employ, through priesthoods, cults, or secret societies.
12. And not touched upon thus far, but prevalent in most underworld traditions, is the strange fact that they are often accompanied by unpleasant or noxious odors; described as "sulphur and brimstone," "sulphurous," or "like rotten eggs." This is the smell of hydrogen sulfide, a very real chemical compound which is found in great quantities deep within the Earth. This will be explored further as well.

(See Color Inserts)

With all of these things in mind, we must now move to a region somewhere in the twilight of folklore, in a gray area which overlaps myth, fiction, and reality, in order to gain a clearer understanding of the inspiration for, or result of, the archetypes which have been clarified to this point.

SECTION THREE: MODERN FORTEAN AND CRYPTOZOOLOGICAL CONNECTIONS

With the advent of the "scientific century," i.e., the twentieth, occurrences which would have in earlier times been deemed "magical," demonic, or the work of faeries, gods, or witches, were given descriptions or names which downplayed their inexplicable nature and sought to somehow fit them into a comprehensible, "modern" worldview. The problem was then, and continues at present to be, that such anomalous phenomena have defied the most rational, skeptical, and "scientific" explanations. Also termed "Fortean" events, after anomaly researcher and writer, Charles Fort, these occurrences have continually thrown a philosophical monkey-wrench into the cherished machinery of scientific dogma. In other words, science has no explanation for these events or discoveries, so it largely ignores, ridicules, or dismisses them out of hand.

There is no doubt that many of these phenomena are genuine, and are for now beyond the ability of the currently dominant belief-system, Science, to explain. This brings to the fore the first category of research which may tie in very closely with "what has gone before."

PART ONE: CRYPTIDS AND MYSTERY CREATURES

"Cryptozoology" is a term which was coined by Belgian biologist, Bernard Heuvelmans, in order to designate a field of study which is concerned with unknown creatures or animals, which are also referred to as "Cryptids" ("hidden" animals). For the purposes of this study, the term cryptid will not be used to designate known or suspected known animals which are simply "out of place" or out of their customary environment. It will be reserved for those creatures which are truly "unknown."

The comings and goings of such creatures, their sudden appearances in remote, rural, or densely-populated areas, and their equally-sudden disappearances with scarcely a trace left behind, have long befuddled researchers in the field of cryptozoology. With a careful examination of specific unknown animals or entities and their habitats, interesting connections to the ancient traditions are revealed.

1. Lake Monsters and Serpents.

Many freshwater lakes around the world have long-standing traditions of "lake monsters" of various types, ranging from humped to sinuous, scaled or plated, to sporting "elephant-like" hide and "horselike" manes of hair. Most such creatures appear sporadically, often in high mountain lakes or large isolated lakes of great depth, with little or no access to the sea. Questions which have yet to be answered include:

1. How do such creatures maintain a breeding population of sufficient numbers to provide genetic diversity in relatively closed environments?
2. How do these populations manage to both feed themselves and maintain a high core temperature, their bodies being of immense size, and in usually cold waters with limited food resources?
3. What is the relationship of "lake monsters" to "sea serpents?"
4. If migratory, how do large populations of such creatures travel from lake to lake, or from lake to ocean, and ocean to lake?

Without a doubt, the most well-known of such beasts is the "monster" of Loch Ness, a large, deep lake in the Highlands of Scotland. Loch Ness is connected to the sea by the River Ness, yet sightings have only very rarely taken place in or at the river. The loch itself, like many in Scotland, Ireland, Scandinavia, and North America, has a long-standing tradition and reputation for being haunted by "kelpies," or shape-shifting., horse-headed, serpentine entities. These creatures are often described as having horse-like heads and manes, long serpentine necks, and huge, humped bodies with four flippers. Given the frigid water conditions at the loch (42 degrees Fahrenheit, on average), and the mammalian characteristics of the creatures, it stands to reason that they are either mammals, or some sort of endothermic (warm-blooded) plesiosaur descendants, or dinosaur descendants (many of the theropod dinosaurs were endothermic, as are their descendants, the birds). Given the additional fact that they sometimes sport horns or "ossicorns" like those of a giraffe, the mammalian hypothesis (descendants of zeuglodons, which were primitive whales?) may be the best bet: perhaps a range of animals with similar habits are represented.

The questions of where they originate, where they go when they vanish, and how they maintain a stable diet, still remain. A look at the structure of Loch Ness provides some clues.

It has been determined that at some point in the distant past, Loch Ness was an arm of the sea. Now, it is seldom less than 300 feet deep, and routinely surpasses 700 feet in depth. The bottom of the lake runs up in many instances to meet underwater cliffs, and the submerged sides of the mountains which surround the loch. Sonar has indicated that the bases of these cliffs are honeycombed in places with very large caverns, which are of undetermined depth or extent. Could these lead to even more vast underground caverns under the mountains, some containing air and eventually linking to other lakes and to the sea? It would certainly provide a feasible explanation for the mysterious movements and disappearances of these creatures, and perhaps to the question of their diet, which may be also that of an ocean-going animal, or "sea-serpent."

Perhaps the lakes and lochs which boast of such creatures, around the world, are simply places to breed and spawn in relative safety from the large predators of the oceans. When it is taken into account that many other "vortex-haunted" lakes around the world have been found to have deep caverns which account for their "vortices" (created by the suction of even larger, water-filled caverns, and the movement of huge volumes of water), the whole matter shifts into perspective. Such vortex-sporting, cavern-hiding lakes include Lake St. Jean and Blue Sea Lake in Canada (the latter home to Misiganebic, a "horse-headed" dragon), as well as Pohenegamook, Massawippa, Memphremagog, and on and on. Loch Ness and the Walchensee in Bavaria, among others, are said to connect to the sea in similar fashion. Given the distance of some of these bodies of water from the ocean, it is not unreasonable to assume that many deep aquifers are actually tributaries or parts of vast, subterranean oceans of freshwater, which may now be in danger of depletion by human usage from the surface.

Is there a connection between the reptilian-yet-mammalian "dragons" of large cold lakes, and vast cavern systems that may lie beneath them? The evidence would seem to indicate that this is the case. The cryptid dragon, with his underground lairs and underwater entrances to them, is apparently as active and secretive as the Chinese and Japanese versions ever were.

2. Hairy Humanoids.

Many eyewitness accounts of hair-covered humanoids have been reported and documented over the years, by people from a wide range of backgrounds and professions. While Bigfoot or Sasquatch is the type whose description is known to most people, perhaps equalled by the Yeti of the Himalayas (the "Abominable Snowman"), the literature would indicate that not only are these living, flesh-and-blood creatures, but they come in a wide variety of physical configurations. Yet like the lake monsters, they seem to "pop in and out" of our surface world reality, and are amazingly difficult to track for creatures so large.

In addition to the towering, hirsute bigfoot form, witnesses have reported stunted, clawed versions with savage dispositions; black, "winged" varieties (the "ixals" of Mexico); screeching horrors with shaggy, hair-covered faces; gigantic varieties which would dwarf Sasquatch (like the Yeren of China), and more. In addition to the "rotten egg" or "skunk-like" odor; which often accompanies most of these forms (as it does both Bigfoot and the Shampe of the Choctaw Indians, or the Florida "skunk ape"), many sightings and encounters have taken place in or near caves or cavernous regions, and abandoned mines and other man-made tunnels or potential entrances to unknown, underground areas. Consider the underworld-exploring Enkidu, the hell-plumbing Chinese "Monkey," hairy Scandinavian trolls, and the fanged, baboon- or dog-headed monsters of Sumer, Egypt and other cultures, and a connection again becomes clear.

These creatures come out into our world, usually by night, and then retreat into an "invisible" cavern or subterranean world about which only they know, vanishing "without a trace." This does not have a connotation of the supernatural, but has the connotation of a natural phenomenon, and is a logical conclusion. Not only does this indicate an underworld connection, but it also indicates an intelligent "covering of tracks" and of evidence which is beyond that of ordinary animals, and can only be equated to the premeditated and well-planned actions of human beings--or of something at least as clever. It also begs the question of, at some level, possible human involvement of a covert nature, in not only "hairy humanoid" matters, but in other, more bizarre, cryptid accounts. And disinformation--in the form of ridicule, evidence-tampering, or intentional hoaxing--may also come into play at times. If this does occur, it does not automatically disqualify the phenonmenon as a genuine one. It only complicates matters by introducing what may be the intended element into the equation, that element being obfuscation.

3. El Chupacabras and Other "Hybrid" Beings.

Puerto Rico is a small island, only about 150 miles long by 45 miles wide; yet it has experienced enough paranormal and cryptid activity in its history to be a continent. The native Taino, an Arawak tribe of the West Indies, feared creatures like the Araidai (jungle goblin) and Konokokuyuha, a type of evil dwarf. The island has long been a haunt of hairy humanoids, mysterious winged creatures (some described as "pterodactyls"), and Unidentified Flying Objects, with the entire human populations in some areas at times living in terror.

During 1975, Puerto Rico was stricken by a mysterious wave of livestock mutilations, accompanied by UFO sightings. Many times animals were drained of blood and left like emptied wine-skins for investigators to puzzle over. Twenty years later, in 1995, things turned really weird.

The mutilations and blood-drainings returned with a vengeance, and over a two-year period, the "culprit" was seen or encountered first-hand by dozens of witnesses. Termed "El Chupacabras," or "The Goat-Sucker," the creature seemed like something out of a cartoonist's nightmare: Described as four to five feet tall, generally reptilian in appearance, with a kangaroo-like hip- or haunch-structure, spines along a back-ridge, and a humanoid head with protruding, incandescent eyes reminiscent of descriptions of the "gray aliens" of UFOlogy. Sometimes the creatures would be gray-green and hairless, sometimes they would have flat black or gray fur. With folding flaps of skin under their arms which connected to their ribcages, they could purportedly glide like flying squirrels, or fly outright, and they had three-pronged, retractable mouth-organs used for slicing, dicing, and sucking blood and fluids. At first it was thought that the sightings were surely an elaborate hoax, or simply hysteria, until the sheer number of incidents and reports reached overwhelming proportions. Hair was obtained from one mutilation site, and analyzed by a Japanese lab; while deemed to be "similar to the hair of a wolf," it was labeled as coming from an animal which was "unknown."

(See Color Inserts)

The creature was a typical cryptid in some respects, seemingly popping in and out of existence at a moment's notice, and generating fear and confusion all over the island. From Puerto Rico, the Chupacabras events quickly spread to the rest of Latin America, including Mexico, South Florida, and the desert southwest of the United States. This marked it as a distinctly cultural phenomenon (Hispanic), but that by no means indicates that it was not genuinely taking place. A connection was soon suspected in Puerto Rico between El Chupacabras and the mountainous, rainforested region of

El Yunque, which is honey-combed with mostly-unexplored cavern systems. In fact, caverns underlie much of the island. Several witnesses reported seeing the creature or creatures fleeing to the El Yunque region, when discovered or pursued. Soon, other parts of the Spanish-speaking world were reaching similar conclusions. And, like the demons of old, an acrid, debilitating foul odor often accompanied the visits of the beast.

One brave soul, Jesus Sanchez, was victimized repeatedly, his rabbits being drained of blood. Finally, at four in the morning, he surprised the invader by blinding it with a bright light. When it rushed to escape, he struck it with a machete, only to discover that "the blow sounded as if it had hit a drum." The creature, unscathed, made its disappearance as usual.

What is to be made of a report like this? How can a creature both "have hair," yet "sound like a drum" when struck (or shot, for that matter), and escape unharmed? Why the almost methodical and indeed selective nature of the animal mutilations? Sometimes only blood was taken; other times, specific organs like livers would be missing. It should be noted that blood and the "meaty organs" were the very items offered to the dark gods of antiquity, in the ancient Middle East and even in Pre-Columbian America. The description of El Chupacabras is very reminiscent of that of the Sumerian Utukku, or even of the Gala, and of the Egyptian Ushabtiu, designated as "artificial" lifeforms (yet no less alive) created to do the bidding of the lords of the underworld. Maybe times are tough in the subterranean realms, and the "gods" have taken to stealing the substances which are no longer offered to them; or could it be that the Nagas are coming out to play?

The Utukku are reptilian in aspect, vaguely humanoid, and slightly "winged," just like El Chupacabras

Again we are reminded of the chimerical, genetic hybrids mentioned in the books of Enoch, which were created by the Nephilim (the Sumerian Annunaki and Igigi) before the deluge, when they "sinned against birds, wild animals, reptiles, and fish," creating hybrid creatures to serve them. These monstrosities also terrorized mankind--as the people of Puerto Rico might verify! Perhaps, instead of "looking skyward" for "alien invaders," we should start looking downward, for ancient parasites which are more or less native to our planet.

4. Other Forms.

Creatures similar to El Chupacabras have been seen around the world, ranging in description from reptilian humanoids (the "Loveland Frog," "Lizard Man," etc.); "little people" of a wide variety of descriptions; bulbous-headed and bug-eyed, stick-thin gray humanoids; and winged anthropomorphic forms. Such appearances, or periods of intense "cryptid" or "alien" activity, are often hallmarked by animal mutilations which involve precise (even surgically-precise), selective organ removals and blood-draining--or else mass disappearances of pets and livestock. These sightings cannot all be dismissed as hallucinations or hoaxes, particularly the sighting reports which consist of several witnesses' testimony. The sheer number of sightings is too great to be so easily dismissed, and the number of unexplained animal mutilations and disappearances which correspond with these "flaps" would seem to support this. Sometimes these creatures are confronted with violent force, yet almost always they resound "as if hollow," and are otherwise impervious to human weaponry, as in the case of the "Hopkinsville Goblins" of Kentucky, which were shot by the witnesses. "Mothman" of West Virginia, a tall, winged humanoid who haunted the region of Point Pleasant, particularly an old military storage area which had "underground bunkers," made mechanical, whirring sounds as it flew, and did not flap its wings at all. The now-ubiquitous "gray aliens" have been described as "mechanical" in their motions, moving "stiffly" or "jerkily." This will be examined again shortly, but all of these beg an obvious question.

Are all of these "creatures" simply different makes and models of "utukku" or "galatur?" Are they the "golems" and the "ushabtiu" of old, and are they completely mechanical and artificially-intelligent--or are they a melding of living genetics and non-living technology? Human beings are only now starting to create such technologies, but they do exist. What technology might be amassed through a simple cumulative process by an earlier, prehuman civilized species, with hundreds of thousands, or even millions of years of history behind it?

The entity known as Mothman, whose reign of terror lasted for thirteen months during 1967 and 1968 (as chronicled so well by John A. Keel, in The Mothman Prophecies), made his visits to West Virginia, which it should be noted is one of the most densely-cavernous areas of North America. So is Kentucky, for that matter, where the Hopkinsville "UFO" entities were encountered. The same can be said of many areas of New England, which as a region has been the haunt of "the Dover Demon, (a term coined by cryptozoologist Loren Coleman)," "lizard men," hairy humanoids, "mad gassers" of reptilian aspect, and a host of other creepy visitors. The New England region is also the location of the "Morehedmoodus" area, near East Haddam, Connecticut; in this area the ground shakes and roars within, as if with "underground tempests," or as if titanic subterranean machinery is at work. This noisy and disconcerting phenomenon is known as "the Moodus Noises," and it has yet to be explained by mainstream science, government, or anyone else with a room-full of Ph.Ds. And of course, neither West Virginia nor New England are excessively far removed from another region which is famed for a distinctive "cryptid," "demon," or "creature:" between both is the Pine Barrens region of New Jersey, home of the infamous "Jersey Devil." This creature not only shares some of the physical characteristics of El Chupacabras and Mothman, but in many accounts resembles the winged reptilian forms of the Sumerian utukku, a dragon, or a medieval cockatrice or basilisk.

With their elusive dispositions, similar characteristics (among their "group," as well as with ancient forms), and nasty habit of collecting organs, blood, or genetic material--not to mention a general aversion to bright daylight or other bright lights, more often than not—the cryptid and "monster" scenario is brought into sharper focus. This multitude of beings is not from "elsewhere;" the odds for that are simply too great, given the variety of forms encountered, and taking into account the already-considerable genetic diversity that we know about which has sprung from the biosphere of the Earth. The scientific principle known as "Occam's Razor" would indicate that, given the choice between a staggering variety of beings from a staggering variety of worlds lost in interstellar space, or a naturally (or even artificially) diverse variety of lifeforms which have sprung from one ecological system, (hence, for instance, the mostly-bipedal, two-handed, two-legged, two-winged, four-flippered configurations, all variations on the earthly vertebrate template), the latter would have to be the most rational and logical choice for a theory of origin. No, they are coming and going at will from some place very near at hand, and the circumstantial evidence would seem to indicate that place to be the hidden bowels of our own planet.

PART TWO: RE-INVENTION, PARANOIA, AND DECEPTION

1. THE DEGENERATES BELOW.

In December, 1943, the pulp magazine Amazing Stories opened a can of worms that would eventually grow into a controversial storm of accusations, ridicule, denial, and resentment. An unknown welder named Richard S. Shaver told the world a story which sounded like a blend of paranoid-schizophrenic delusions, hallucinations, and bad science fiction. Working with the "assistance" of Ray Palmer, Amazing Stories' editor, Shaver began to tell his allegedly "true" tales of deep cavern-civilizations, genetically-degenerated, technologically-advanced troglodytes, and an ancient high-technology which gave these beings amazing powers over the unsuspecting inhabitants of the surface world.

Strangely, in addition to their "death-ray (laser or electromagnetic weapon)," "thought-ray (telepathy)," and illusion (holographic) "mech," or machines, these new incarnations of the "deep dweller" concept also had silent, flying ships which Shaver described as "disks," which were kept hangared beneath the Earth's surface, and were used regularly. This was five years before Kenneth Arnold's famous "first" sighting of nine objects over Mt. Ranier in Washington State, objects which he referred to "saucer-like," hence inadvertently helping to coin the modern term "flying saucer." Additionally, the troglodytes could produce "solid illusions" of creatures or objects, even of their disk-like craft, which were temporarily physical, then dissipated or were "turned off." How many times have UFOs just "disappeared" from radar screens around the world?

Richard S. Shaver and one of his "picture rocks."

Shaver's underworld was inhabited by two ancient races which had originally been one, which he termed "dero," for "detrimental robot ("degenerate" would have been more accurate)," and "tero," for "integrative energy robot (the "good guys")." According to the welder-turned-writer, he had not only been contacted and tormented by the beams of these beings, but he had been "helped" by the tero in defeating the assaults of the dero, and had actually been in the cavern-world himself. Readers responded to his accounts in two ways: they either reviled and ridiculed him and his "underworld," which was often depicted as sexually and violently brutal; or they wrote in with letters of support and accounts of their own, in an effort to corroborate his stories! Ultimately, the "Shaver Mystery," as it was called, drove Ray Palmer from Amazing Stories and into business for himself, and generated an uproar (and an income) for both he and Shaver that would last, to some degree, until both of their deaths in the 1970s.

One of Richard Shaver's "Picture Rocks," cut with a rock-saw,and revealing human women fighting a reptilian humanoid similar in form to "El Chupacabras."

(See Color Inserts)

Was Richard Shaver unbalanced and delusional, or was he simply a talented and outrageous storyteller? Or was he telling stories which were half-truth, half-confabulation? In his youth he had spent some time in a mental hospital, which generated contempt and ridicule from others throughout his life; but would not actual, repeated contact by such beings as his "dero" push even a strong mind over the edge and into partial delusion? "Confabulation" is one common reaction which the human mind uses as a defense mechanism, when it is hiding a traumatic event from full recollection.

Regardless of the proportion of truth to fantasy in his accounts, Shaver's choice of words to describe his tormentors--"robots--" is interesting. While "fleshly" and possessed of animalistic lusts, the dero and tero are nevertheless "programmed," victims to a greater or lesser degree of "harmful" or "de" (destructive, degenerate, etc., according to Shaver) thinking. He attributed the degeneration of the subterraneans to damaging solarand cosmic radiation (from which they still hide underground), and to the destructive radiations of their own "stim" (sexual and other stimulation) machines. They kidnap surface humans, particularly women, for sexual pleasure (or procreation?), eat human beings, and take great delight in causing mayhem, destruction, confusion, and terrifying apparitions (holograms) in the surface world.

As an aside and interjection, Shaver's claims of the abilities of dero machinery would explain many additional mysteries which have not been examined here so far. To touch on just one example would be to examine the case of the Bell Witch of Robertson County, Tennessee, a malevolent and violent entity which terrorized a family in the early 1800s. The "Bell Witch," so-called because "she" haunted the Bell family, went to great lengths to create the illusion that she was omnipresent and omniscient, although she was fooled and eluded upon several occasions by her victims At first beginning with standard "poltergeist" activities, the "spirit" worked its way up to invisibly-dealt physical blows and attacks, and throughout the ordeal which spanned many years, spoke to, sang for, and reviled the Bell family. The activities of this entity were witnesses by hundreds of people, including Andrew Jackson, the seventh President of the United States. Only now, in our "modern" age, is it noticed that at first the "voice" of the "witch" was booming, hollow and metallic, only later and after considerable practice taking on a regular, well-modulated tone; and the origin of the thing was said to be in the "Bell Witch Cave," a deep and mostly uncharted cavern system that was at that time on the Bell family property. The "Witch" itself claimed that the cave was its home, and forbade anyone to enter there. Local Native Americans had shunned the cave and surrounding land as haunted or cursed, long before the first white settlers arrived. To this day, witnesses swear that the still-unplumbed cavern system is haunted and even guarded by the same evil being.

A terrible story, indeed, and sure to engender fear and foreboding--perhaps the perfect type of story to make certain that a short and direct route to some underworld lair, loaded with "special effects" equipment or "mech," is not tried or violated....Spelunkers and other brave souls have yet to seriously attempt to explore the deep cavern system called "The Bell Witch Cave." It looks like the project may have achieved its desired goal of keeping people out, after all.

As evidence of his obsession with proving the veracity of his stories, Richard S. Shaver was one of the few "contactees" by "alien" or nonhuman forces to make an effort to produce physical evidence. He claimed that many rocks and boulders in specific regions around the world are actually great, shattered libraries of ancient crystals, containing three-dimensional images, or hologram-records. To prove this point, in later years he took to finding such "picture rocks" and cutting them open with a rock-saw, to reveal what he said were glimpses of the ancient records. He would use paint to accentuate or "bring out" these images, but only worked along the edges or strata of shapes which were pre-existing. Some of his "picture rocks" were in fact quite startling, depicting humans or human-like beings, human-animal hybrids, giants, and humans battling with creatures which look suspiciously like Nagas, El Chupacabras, or something similar. In 1975, Ray Palmer featured many of these pictures, along with Shaver's commentaries, in The Secret World.

(See Color Inserts)

The forms described by Shaver should be familiar by now. The dero--dark elves, or trolls, or Nagas, or utukku/ushabtiu--and the tero--the "noble faeries," Tuatha de Danaan, or " Aryan/Nordic Masters--" have simply manifested in a more contemporary, pseudo-scientific form. The naga "vimana" disk-craft are present, and the abduction and genetic rape motif of all underworld goblins/faeries/ Nagas is present as well. The "treasure of the Nibelungs," or the hoarded wealth of dragons, has been replaced by ancient and super-scientific "mech," a technology which bestows godlike powers upon its owners. And like the trolls, goblins, vampires, and the rest of the ages-old underworld crew, these subterranean beings are in hiding from the sun, living in mole-like terror of its direct rays. A relationship between Shaver's underworld inhabitants, and those of the past, is hopefully not so much literal and based in some terrible reality, as it is archetypal and obvious.

2. THE MASKS MAY CHANGE, THE PLAY'S THE SAME

The final three decades of the twentieth century have been filled with rumors: conspiracy theories about alien abductions, plots for world domination shared between aliens and "secret government," illuminati cabals, and the like. Several interesting variations on age-old theories have permeated the field of UFOlogy and conspiracy literature alike, and in many places the two have overlapped and in fact become completely blurred into one. It is these latter areas which are of particular interest and significance to the study at hand.

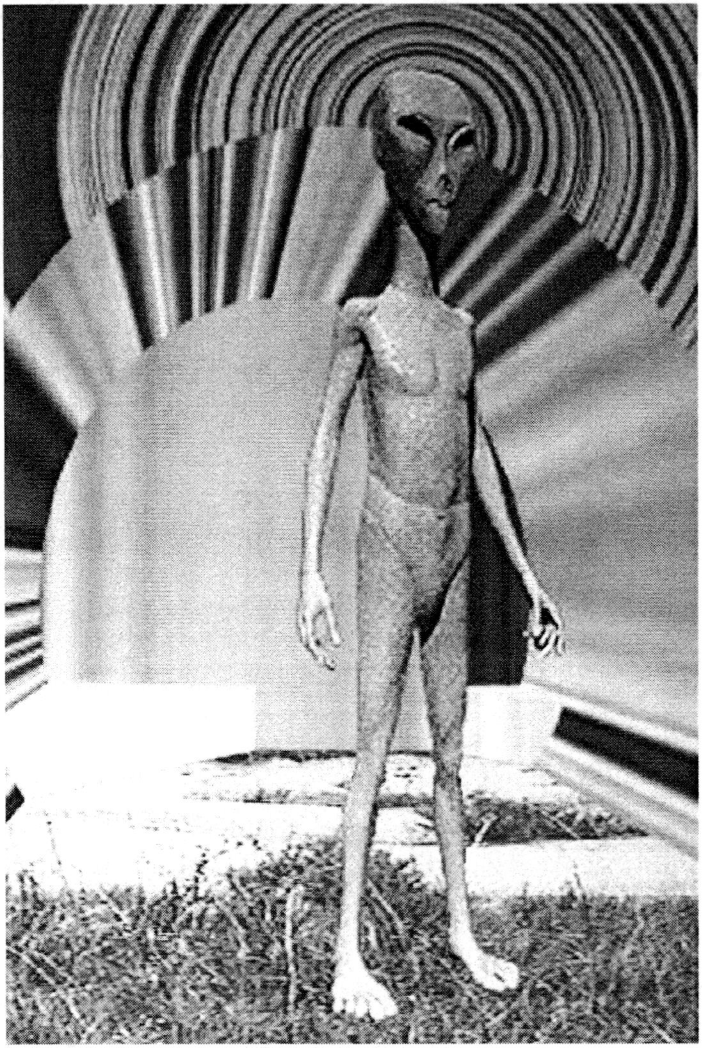

(See Color Inserts)

One of the most commonly-reported manifestations of "alien invaders" are the "ebe (extraterrestrial biological entities)" type, commonly called the "grays." These beings are the variety most-often reported in abduction accounts; and they are highly ectomorphic, with scrawny, underdeveloped limbs and bodies, oversized heads, and extraordinarily-large, black eyes (some abductees have reported these to be actually "reptilian" or "birdlike" eyes with slit pupils, the "blackness" only a protective artificial film, like sunglasses). They are from three to five feet in height, averaging around four feet tall. Again, their very physiology gives away their origins, for large, protruding eyes, with large slit pupils and needing an artificial protective covering, would be hard to equate with a race which has mastered, and perhaps been genetically-prepared for, interplanetary or interstellar travel.

Outer Space is an extremely bright, radiation-filled environment. The type of eyes described are those of a creature which spends most of its time in the darkness, as they are designed for optimum light-reception; and the shaded coverings, for venturing out into the surface world, are really self-explanatory. Similarly, their bodies, small and easily-maneuverable through tight spaces, with small surface area and a minimum of body-weight, are ideal for an underground environment. Their method of locomotion, generally described as "shuffling," and "hips moving strangely" or "sideways," is another indication that they have developed in a relatively cramped place, or even one where tunnels are commonplace. They are often described as smelling "musty," "like a snake," or "like rotten eggs."

This type of entity allegedly abducts unfortunate human beings, conducts medical or genetic testing on them, sometime removing sperm and ova for use in "hybridization experiments," or for purposes unknown. The abductions almost always take place under cover of darkness (doubly-helpful to the molesters, since the "harmful" sun has set, and people are drowsy, or asleep).

The victim is taken into a UFO (usually disk-shaped, or shaped like a child's spinning-top) for experimentation, and also for a form of indoctrination, consisting of intense three-dimensional audio-visual presentations. The primary message of these presentations seems to deal with a concern over human destruction of the Earth's biosphere and biodiversity (recall the recurrent "faerie" interest in the same thing—a steady supply of genetic materials).

This becomes even more telling when it is recalled that the first underground atomic tests were concurrent with the resurgence of "witnessed activity" of UFOs on a previously unprecedented scale--Those tests may have rattled more than a few windows in the desert! During the medical aspect of the abduction, local anesthetics are seldom if ever used, and the victim is left with a type of post-traumatic stress syndrome which is at first marked by amnesia and mental distress or unease, sometimes followed by nightmares, self-imposed social isolation, and eventually a total recall of the experience, either through natural recovery of memory, or through hypnotic regression therapy.

Other common themes emerge as well. The entities are described by their victims as "drone-like," "robotic," "clinical," and so forth, and are also described as "reptilian," "lizard-like," or as having reptilian, birdlike, or amphibian skin texture. In many instances the abduction experience moves to a "cavern city," "cavern world," or "underground base."

It is in these latter, subterranean places, that "hybrid beings," having "alien," human, or other animal characteristics are reported as being seen in development, along with cannibalism and torture of human beings. One such site has been alleged by several researchers and "abductees" as being beneath Archuleta Mesa, near Dulce, New Mexico. In these prolonged episodes, other beings enter the scene, such as "hybrid" children who are frail, pale, and ectomorphic, but generally human in appearance. Some abductees have even said that these children look "like fairies!"

Other entities are not so harmless, such as a tall, "human" type with an aristocratic, "Aryan" look; these are generally referred to as "Nordics" in the UFO literature, and bear more than a slight resemblance to the "Light Elves" and "Tuatha de Danaan" of old. Are these the "fairy children," hybrids and "changelings" all grown up and hard at work? Another type of entity is more sinister in both appearance and attitude, described as a "reptilian humanoid," "lizard man," or "reptoid," and ranging from five, to eight or nine feet in height. These "reptoids" are usually characterized as being "in charge" of the other types, but upon occasion are said to report to taller, even stranger entities which resemble skeletally-thin, giant "grays," or even "giant mantids." Normal surface humans, paramilitary in nature or appearance, are also seen in these underground areas.

Another interesting factor is that all of the entities described go to great lengths to convince their captives that they have come from "far away," from distant stars, planets, and "vibrational frequencies" or dimensions. They have come all this distance because they are "worried about humanity," or some such heartfelt propaganda (as they stick a huge needle in a woman's abdomen, or up someone's nose, without even local anesthetic). Yet the biology of all the different types or castes of abductors, as horrific as it might appear to superstitious human eyes, is essentially that of animal forms which are natural to the Earth: mammalian and reptilian. Obviously, it is very important to these beings that such a logical connection not be made. If there is even a shred of truth to UFO abduction accounts, then it is more than apparent that the abductors want their victims, and humanity at large, to believe that they are from "somewhere else." While humanity looks continually upward at enigmas in the sky, what is transpiring beneath our very feet?

Additional folk and literature parallels are apparent in the accounts described. The "grays" are identical to the order-following, human-abducting, drone-like GALATUR, and Ushabtiu of the Sumerian and Egyptian underworld mythologies. In Shetland Island folklore, "little men" who abducted people were sometimes referred to as the "gray neighbors," and the "grays" also bear a strong resemblance in head and torso structure to the cryptid "El Chupacabras." The methodical imps and djinn which served Satan and Shaitan/Iblis come to mind as well, and of course, Richard Shaver's "dero" are similar, especially in their use of high technology. The "Nordics," sometimes seemingly the "enemies" of the gray and reptoid types, are often reported as working side-by-side with them in the underground facilities or labs, which recalls the fact that the some Nagas were said to look "almost human," as well as the mercurial dispositions of the aristocratic or " Aryan" faeries, the light elves/de Danaan. Additionally, the apocryphal hybrid offspring of Nephilim and humans were described as having an extremely "nordic" or Aryan appearance, as described in The Book of Lamech, and the Slavonic Book of Enoch, and were also said to have a reptilian patch of skin ("badge of priesthood") on their chests or elsewhere. In Celtic Welsh myth, the lord of the underworld of Annwn had a "magical cauldron" which produced an endless supply of warriors for him...Was this an "earth-mother, primal-womb" archetype, or was it instead synonymous with producing offspring through gene-splicing, the "cauldron" actually a "test-tube?"

The reptoids or lizard-men are familiar, as Nagas, utukku, ammut, "dragon kings," goblins, trolls, and so on. They, also, are often described as smelling "like rotten eggs" or like sulphur.

(See Color Inserts)

Both the reptilians and the grays have in recent years passed from the UFO research realm and into "conspiracy" literature, where they are said to be involved in "controlling the Earth," or are in the process of "taking over." Several authors have published the theory that the reptilians are masters of illusion, holographic projection, or are in fact physical shapeshifters. which are replacing world leaders, government officials, and public figures, as an insidious "fifth column." Here again are the "changelings" of European folklore, the hologram-utilizing dero, and the "Serpent Men," precisely as described by the fiction writer Robert E. Howard in his pulp-fiction tales of King Kull.

This is by no means a "new idea," revelation, or suspicion, but is as ancient as the concept of an underworld itself.

3. INTO THE OUT OF

This treatise would not be complete without an examination of another mysterious player from the depths, the MIB or Men In Black. These cool cats figure prominently in both UFO and conspiracy literature. Thanks to the excellent work of researchers and writers John Keel and the late Jim Keith, as well as to the work of a host of others, many small and seemingly inconsequential facts have been obtained from witnesses and preserved.

"Men in Black" seem to fall into two categories: the standard men in black, whom may be actual agents of secret government or military investigative/disinformation groups; and the MIB, whom Keel indicates are somehow in league with, or originating from, the unseen forces behind the various cryptid or creature sightings, as well as the UFO phenomenon. It is the latter category, the MIB, which has bearing here.

These "mystery men" are usually described as of varying height and build, but more often than not they are on the thin side; they generally wear dark or black business-like clothing, dark hats, and sunglasses (again, the eye protection from the sun!). They are either "olive-skinned" and vaguely "oriental" or "Asiatic;" or they are "Nordic" or "Scandinavian" in complexion, hair color, and physique. From the basic template of these two forms, strange variations have been reported, such as total hairlessness (not even having eyebrows or eyelashes); overly-large, protruding eyes (a "non-mammalian" trait, for the most part, or perhaps due to an unaccustomed lack of atmospheric pressure); wheezing and other trouble breathing, as if unused to earthly (or surface?) air pressure (aeroembolism); unnatural joint movement and locomotion; "reptilian" or "frog-like" cast of skin texture and facial features; webbed fingers; sulphurous or "metallic" odor; and a host of additional oddities of physical configuration.

Add to all of this the fact that they are often ignorant or in amazement of the most ordinary surface-world activities--trying to drink gelatin, refusing food and taking a pill instead, stealing or asking for small common objects (like writing pens) as apparently prized souvenirs--and they become even less human through their behavior. They often exhibit a strong interest in the sexuality or sexual habits of those they confront, and the longer they remain on the surface, the more erratic, disoriented, lethargic, or "drunk" they seem to become (a side-effect of rapid de-pressurization, or aeroembolism). With cryptic statements they have sometimes identified themselves to their unwilling visitants as being citizens of "the Nation of the Third Eye," an occult or secret society reference still widely utilized in symbology and ritual today by some "brotherhoods" of "enlightened" human beings, but which hearkens back to the "third eye" of the Naga, or the "skull-pearl" of Chinese dragons. Any relationship this "Nation of the Third Eye" might have to modern-day secret societies is debatable, but the parallel exists.

Another clue lies in the fact that they arrive to their victim's doorsteps in shiny, black cars which are in pristine condition--i.e., "like new." The puzzling thing is that these vehicles are almost always decades out of date, and sometimes seem to be composites of several different makes of cars (still out of date). To remain unravaged by the passage of time--oxidation, dry-rot of tires, and so forth--they would have to be warehoused and maintained in an environment of constant temperature and low humidity, away from weather, sun, and extreme changes in temperature. Add to this the manner in which both the cars and the men suddenly and inexplicably vanish, as if swallowed up by the Earth--but by now the premise is obvious. Some caverns are damp, but many others are exceedingly dry, and all remain constant in temperature, year 'round, after a certain depth.

History and folklore both have parallels. Folklore is filled with "dark men," "men dressed in black," and "grim reapers," often identified in previous centuries as sorcerers, demons, warlocks, or other "servants of the devil," and of course the odor; of sulphur and brimstone was a trademark of his from way back. During the plague years of the middle ages, entities resembling both MIB and the now-standard "gray aliens" were often seen in areas which would shortly thereafter be stricken with an outbreak of the dread disease. Throughout history, on the eve of major events, people have repeatedly seen or been harassed by such beings. As Keel points out in Disneyland of the Gods, and his other excellent books, Julius Caesar, Napoleon, and even Malcolm X all reported encounters with this variety of terrifying being. Hitler also was alleged to have had his share of midnight visits from a mysterious "Tibetan," and through him to have met "the New Man," a sort of super- Aryan, whom he believed came from the interior of the Earth, and of whom he was most afraid.

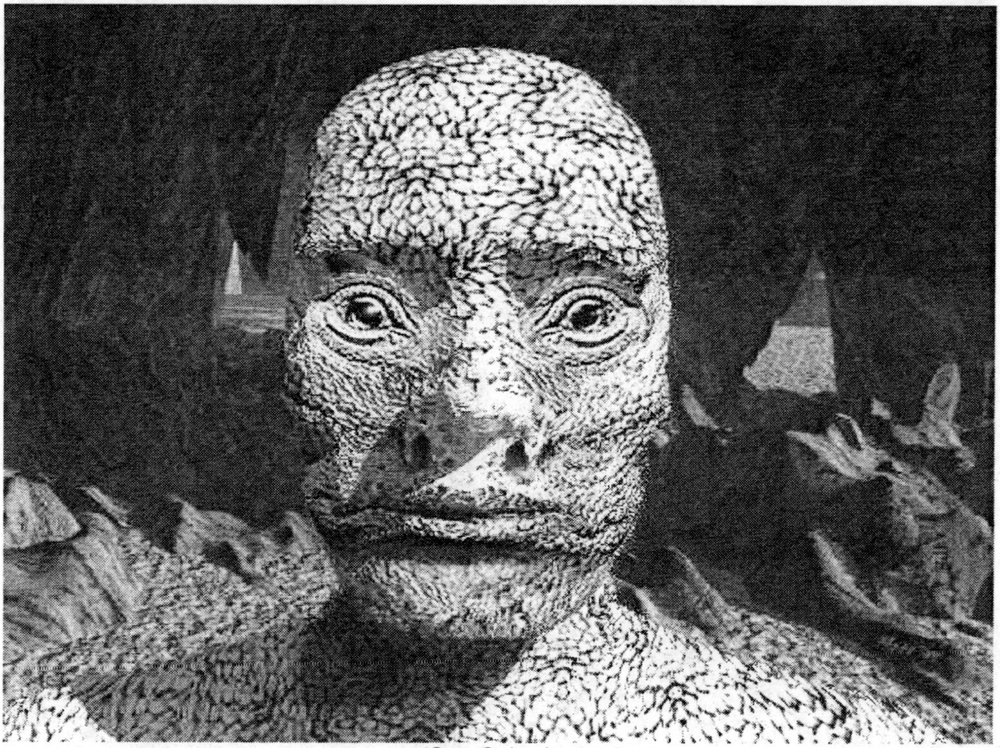

(See Color Inserts)

How much fear, confusion, and human suffering can be traced to uncanny visits from these "robotic yet human" agents provocateur? Where are they from, who do they work for, and what is their long-term agenda? Could it be that their goal is to generate confusion and divisions, to keep humanity "looking upward" for an invasion that will never come? The "invaders" may already be here, and may have been here all along. The evidence is in the folklore, religions, myths, literature, and archetypes of humankind. Perhaps they're not "extraterrestrial," but are "intraterrestrial," and there's a vast, unknown world beneath our feet, stretching down through secret, twisting tunnels and deep caverns to the Mohorovocic layer itself, which is an anomalous cavern region, deep beneath the upper crust. What if there is an unknown world beneath our feet, a world that is dependent upon the biodiversity and genetic wealth on the Earth's surface; a world which has been exploiting that wealth for thousands or millions of years, victimizing the ignorant savages who roam the face of the sunlit world? Or could all of the evidence be circumstantial and without merit, simply a misinterpreted conglomeration of coincidence and mis-identified animals, natural phenomena, and archetypes from the human collective unconscious? The critic could toss in an endless supply of over-active imaginations, down through the millennia, but the evidence—dating back as it does for thousands of years of human traditions, and continuing right up through the present day--says otherwise.

A reptilian humanoid in the underworld.

The reptilian, vampiric, robotic, and demonic are all characteristics which have been attributed to underworld beings down through the ages. They have haunted mankind's imagination and nightmares since our most remote time. These "archetypal" images speak to us of supernatural terrors and mysteries, but what if they are instead natural, just technologically-advanced to a degree that has only until recently been beyond our comprehension--and therefore considered "magical" or impossible in their level of sophistication? If the latter is the case, then we would do well as a species to become more aware of not only our own planet, but of our mythic and folkloric heritage, for it speaks not of a symbiosis, but of a nearly-invisible Parasite, dwelling in the depths.

Two examples of UTUKKU (djinn or genii), dispatched from the KI-GAL to the surface world.

Bartolomeo's view of Dante's Inferno.

A hero fighting a scaled, humanoid creature, similar to a Greek Gorgon.
Note the "third eye" of the creature.

Part II:

HIDDEN NEIGHBORS:

The Genetic Victimization of the Surface World

1. OUR SURFACE FARM, AND PLAYGROUND

(See Color Inserts)

One of the universal, recurring motifs of visitors to the surface world from faerie or fairyland is that of the fairy lover. One of the attributes of the fairy lover has always been illusion, or glamour--illusion and confusion of location, dilation or distortion of time, and most tellingly, illusion of identity. Glamour, in it's original Celtic sense, was the name for the seemingly transformative powers of disguise and illusion possessed and exercised by fairy-folk both benign and malignant, beings of the air, the seas, or the underearth. The willingness of the fairy lover to use this power, selfishly and without compunction, is a worldwide constant in ancient myth and folklore, as well as in modern UFOlogical and other paranormal accounts. Equally as selfish is the fairy willingness to kidnap infants and abduct fertile young people of both sexes, apparently for crossbreeding purposes. In many accounts, the faeries of the British Isles and Scandinavia would often be so bold as to tell their victims that this was the reason they sought human children, promises of marriage, or repeated visits to the terrified or enamored human being, for reproductive purposes and "fresh blood" or genetic material among their own families or tribes.

A record of human relations with a non-human race or species has been available all along, but has been distorted by technologies or powers beyond our understanding--providing only a hazy view as if seen through smoked glass, or as if vaguely glimpsed, creeping up behind one in a warped and illusory funhouse mirror. The enormous amount of first-hand narratives and folk "traditions" have been dismissed all too often as primitive belief systems, superstition, or foolishness. There is a difference between "fairy traditions--" the first-hand, second-hand, or strangely-experienced knowledge and encounters of a tribe or people--and "fairytales," or stories, which mix "fairy traditions" with confabulation, fantasy, oral storytelling (itself a blender of sorts), and morality tales. Upon close examination and comparison, the "traditions" themselves seem to provide a cohesive and revealing framework from which much information can be obtained--information about the habits, appearances, customs, and so forth of non-human, more-or-less hostile, subterranean races. Often this information would find it's way into "fairytales," in order to lend an air of authenticity to the story at hand. Soon the researcher realizes, however, that this wealth of anecdotal evidence has influenced to some degree or other every folklore tradition of the world, and continues to influence our plays, our literature, our art and most other forms of expression. These things that so confound and fascinate us have become some of our most-constant, ever-present archetypes.

The British, Scottish, Welsh, and Irish subterraneans were believed to dwell in a variety of locations which connected directly into a larger, nebulous region, called by a variety of names: Tir Na N Og, Faerie, Fairyland, Elf(Alf)land, and many, many others. To the Norse these regions were known as Alfheim and Svartalfheim, and should not be confused by modern students of folklore with a deeper, darker place of greater terrors, called Nifleheim. Entrances to the regions inhabited by these beings could be found in raths (ancient hill-forts), burial-mounds (barrows, sitheans, knowes, brochs, duns, and so forth) and other ancient earth-works, caverns or clefts in mountains and hills, and at the bottom of deep lakes or lochs. Some confusion occurred, largely due to theorizing on the part of both victims and collectors of folktales, that the fairy-folk were fallen angels, nature spirits (elementals), or the hyperactive ghosts of the pagan dead. The problem with these theories is that they are contradictory to the activities described, and the purposes which obviously underlie these activities--that is, the gathering of genetic material, ranging from human and animal to plant samples, and the actual physical interbreeding which has been described. It would seem that "spirits" of any description would have no use for these things. This does not necessarily disqualify the existence or involvement of factors which would be considered "paranormal" or "spiritual" in nature; rather, it points out the consistent evidence that perhaps some aspects of the phenomena described are due to an unknown and non-human technological agency.

Collections of Celtic and Germanic fairylore and sagas are filled with such interactions between human beings and "others." Always, an influx of human genetic material seems to be at the heart of such activity, and the emotion of love seldom, if ever, enters into the equation. From the human point of view, such interaction almost invariably leads to suffering, hardship, and sorrow, and ancient taboos existed to provide warnings and reminders of the harassments from the "serpent (pre-Adamic) races" from below:

> *"And I will put enmity between thee and the woman, and between thy seed and her seed; it shall bruise thy head, and thou shalt bruise his heel." --Genesis 3:15.*

This is believed by theologians to be a reference to the eventual birth, crucifixion and resurrection of Jesus Christ, but it may have an additional meaning--it establishes the belief that humanity dwells "above" another race or species, that of the "serpent." Older versions of the Genesis story and the Fall of Man, from Sumer, leave little doubt that many in the ancient world knew of this "mystery" of old.

In Greece, the philandering Zeus, king of the gods, regularly descended from his palace/grotto on Olympus in order to seduce and produce offspring with mortal women. The God-Mountain, holy mountain or hill is a motif which appears in most mythic traditions, and is obviously associated with both the sky and the power of flight, as well as with the depths of the mysterious and unplumbed Earth. Zeus had his underworld connections, as his brother Pluto/Hades was in charge of the Underworld, and it was Zeus himself who had imprisoned the "fallen ones," the Titans or Elder Gods, there in the deepest and most-inescapable pits. One of Zeus's favorite tricks was to take the form of either a familiar person or an animal during his libidinous assaults, although how a woman can be seduced by a swan is still hard to fathom. At least Europa ran from the bull!

Nimue imprisons Merlin beneath the Earth.
Illustration by: Arthur Rackham
(See Color Inserts)

Similarly the sorcerer Merlin, himself said to be the half-human result of the union of an underworld being and a mortal woman, covered Uther Pendragon with the likeness of his enemy Gorlois, Duke of Cornwall, so that he might impregnate his rival's wife Ygraine and produce the child which would become Arthur, the mythical king of Britain. Uther wanted pleasure; Merlin had an agenda of his own. It's meaningful that Merlin's father was said to have been an incubus, or a demon/underworld entity capable of having intercourse with human women and impregnating them. Merlin had to bring together the desired genetic partners, and this seems to have been simply underworld manipulation of human history and politics as usual, as far as subterranean interference in human affairs go. The fact that Uther and Arthur's standard was a crimson dragon is more meaningful than might at first seem apparent, as shall be seen. Further examination of Arthurian legend, based as it is on much older Celtic myth and archetypes, seems to indicate that it may in fact be a thinly-veiled fable concerning two warring supernatural or underworld forces, one more inimical than the other, vying for control or breeding-rights of the localized human gene-pool.

This battle for dominance is perhaps symbolized by the warring dragons, one red and one white, and the manipulation of Merlin's emotions and lust by Nimue, the Lady of the Lake. Nimue is synonymous with the cavern-dwelling Hindu Naga-folk, who dwell in palaces beneath the floors of deep lakes and oceans, and she is also very similar to the daughter of the Japanese Dragon-King, who falls in love with a human trespasser and removes herself from her undersea cavern-palace, long enough to leave a hybrid child--and perhaps political influence--in the surface world. Nimue represents the ancient matriarchal powers which seem to go hand-in-hand with underworld beings and deities, but there may be a more concrete reason for this recurrence of matriarchal themes, buried in the mists of time (as will be seen later).

As for Merlin, outcast by humans and his own underworld tribe alike, this provides an explanation for his fascination and lust for, and willingness to obey and work with, the Lady of the Lake. Merlin not only had his own subterranean domain, but he was eventually imprisoned there by the treachery of Nimue, once his usefulness at genetic expertise and manipulation had been overshadowed by his attempts to affect events in the human realm.

The King of the Snakes or Serpents is sometime portrayed as humanoid, sometimes as purely serpentine. He holds the key to most secrets and many treasures, technologies, or magic, an din this case stolen goods.
Illustration by: H.J. Ford

The direction of civilization--i.e., "Camelot--" may have taken a better route than it would have otherwise (if the invading "Saxon," or continental, forces had not been routed), but it was still designed to keep interloping genes out of the local pool (that of Britain), and keep the population rich in certain desired characteristics. Nimue's hand-picked (intentionally bred) proxy was on the throne, at least for a while. Arthurian myth and romance, with roots lost in the antiquity of pre-Roman Britain and derived from the Welsh Mabinogion and other Celtic traditions, seems to be desperately trying to burst forth with an ancient and mysterious message which is cloaked in more historically-recent, post-Roman, Christianized symbolic constructs.

The careful monitoring and maintenance of bloodlines or genetic groups seems to have been a major preoccupation of faerie folk and other subterranean beings, as well as of ancient gods and demons. Ireland also is filled with tales of fairy lovers: relationships between "Milesians" or ancient Irish human beings, and the Tuatha de Danaan, who were the underworld gods of old Ireland who have since dwindled into the Gentry, Sidhe (Shee), or Fairies. One encounter of this "pagan reality" was purportedly witnessed by St. Patrick himself, described in "The Colloquy of the Ancients;" while speaking with Caeilte, a member of the Fianna (a legendary group of Irish heroes), Patrick is astonished to observe a beautiful fairy woman coming out of the cave of Cruachan, wearing a green mantle and a golden crown. Although the fairy woman is youthful and beautiful, Caeilte is an ancient and withered old man. When Patrick asks about this, Caeilte answers:

> *"She is of the Tuatha De Danaan who are unfading...and I am of the sons of Mil (human Irish), who are perishable and fade away."*

The question is, of course, at what price to humanity is this "longevity" or youthfulness purchased?

It would seem that it was always in the interest of these subterranean, older races or species to keep a relatively pure genetic pool of human beings near at hand for their own use or convenience. Recent scientific evidence from Ireland, reported by Reuters News Service and the science journal Nature, bears this out.

According to these reports, a group of geneticists at Trinity College in Dublin, led by Daniel Bradley, have determined that the oldest "pure" racial bloodline in Europe still exists in the West of Ireland, the furthermost western edge of Europe. According to the report (March 22, 2000), Bradley and his colleagues have used surnames along with the male Y chromosome in order to reconstruct a 1,000-year-old genetic map of Ireland, demonstrating that the Irish are truly representative of a "race" which is distinct from other Europeans--a race apart.

The Serpent King, hanging out with the genetically-potent victim.
Illustration by: H.J. Ford

Bradley stated to the press that "When you look at this old genetic geography of Ireland what you find is that in the West (of the Emerald Isle) we are almost exclusively of one type of Y chromosome."

The Y chromosome, passed down exclusively from father to son, is an object of interest and study to geneticists, as it accentuates differences between populations. According to Bradley:

> *"It is inherited as a unit so the information you get from it is of a special type."*

The researchers examined the Y chromosomes of men with Gaelic surnames from the western-most province of Connaught, finding that 98.3 percent had a group of genes on the Y chromosome known as haplogroup 1.

Bradley observed, "When you look at Gaelic surnames they are different in frequency of Y chromosome types from non-Gaelic surnames."

Furthermore, more than 98 percent of men with Gaelic names in the western half of the island had haplogroup 1, but numbers dropped significantly on the eastern side of Ireland, whereas there's a gradient of haplogroup 1 across Europe which starts at almost zero in the Far East and grows to almost 100 percent in the west of Ireland. According to Bradley and his team, this gives a snapshot of what the original ancient racial type of Europe once looked like, before successive invasions and influxes by other racial or genetic groups.

Was this what Nimue and Merlin were seeking to protect from "Saxon" contamination? What better place to maintain a genetically-pure control or access group or groups than on two large islands like Britain and Ireland? But the British Isles are not the only group of islands with such a combination of relative isolation, and legends of strange interbreeding activities.

Japanese legends of serpent/dragon and human marriages, seductions, and liaisons abound. Sometimes the offspring of such unions were said to be either more human-like or more reptilian in nature or appearance. Such shapeshifting. dragon-beings, appearing as beautiful men and women, would walk into the life of prominent samurai, warlords, or other notables, obtain the desired genetic influx, and then depart. Occasionally the offspring of such unions would be left in the surface, or human world; the dragon-person would return to his or her abode in caverns beneath the mountains or under the sea.

(See Color Inserts)

One such story is an ancient Japanese tale which supposedly occurred in the village of Siota, at the foot of Mt. Soba, one of Japan's many "holy mountains." A maiden named Mimoto, the daughter of a prominent merchant and civic leader, became enamoured of a handsome and mysterious youth whom she took to meeting in the woods at night. Lured by his flute and seduced by his advances, Mimoto lost her sense of discretion and caused great dismay for her parents, who planned a disruption to her affair with the unknown young man. At the insistence of her parents, she attached a strong red string to the young man's garments. After his departure before dawn, Mimoto, her parents, and several strong men followed the string through the vales and forests, until at last it led them to a cavern on the slopes of Mt. Soba.

A Japanese underworld queen, complete with reptilian dragon servants. Illustration by: Utagawa Kuniyoshi. **A Japanese underworld queen, complete with reptilian dragon servants. Illustration by: Utagawa Kuniyoshi**

At this point in the story, mythic elements of a more symbolic nature take over--the "young man" is heard inside the cave, moaning in pain, and when Mimoto calls to him he informs her that he has "a needle in his throat," presumably the needle which she had used to attach the thread to his garments. After some coaxing, he is persuaded to come out into the light of day, and the villagers of Siota behold his true form: that of a Ryu, or dragon. He is ashamed to show his true appearance, and the people are understandably distressed. The dragon informs his lover that he will die soon, due to the prick of the needle, and that she is also pregnant with his child, whom he apparently can't take with him into "his world" due to his imminent demise.

One interesting point here is that both the bright light of morning, and the "prick" of a needle, proved the undoing of a mighty, scaled, underworld entity. As a parallel of note here, as recently as the nineteenth century traditional Celtic belief in the British Isles held that ironpins and needles, stuck in clothing, blankets, and so forth, would protect people from abduction or from being replaced with "changelings." It might be worthwhile to theorize that this is indicative of a disruption of an electromagnetic field of some sort, resulting in the "illusion" or technology (hologram?) being disabled. But since most liaisons between underworlders and humans take place at night, the radiation of the sun is just as likely to have been the source of the "dragon's" distress as would a "needle" in his throat. Still, the Reverend Robert Kirk, an investigator of the "subterraneans," who died (or, as tradition holds, was "abducted" while investigating the Fairy Hill at Aberfoyle, a natural geological formation) for his nosiness, wrote about preventing kidnapping and seduction by the "subterraneans" in The Secret Commonwealth (originally published in 1691, changed in this work to more contemporary English for ease of reading, and to satisfy those who lay claim to Kirk's own words):

> *"Even to this time the Gaels place bread, the Bible, or something of iron in womens' beds when they are giving birth, to save them from being abducted. And they regularly report that all unclean supernatural beings are terrified by nothing on earth more than they are by cold iron. They (the highlanders) explain that the reason is that Hell, stretched between the frigid winds (of the North) and burning instruments of red-hot metal (the deep interior of the Earth?), as well as the Northern iron (magnetic pole?), causes an attraction to that region, which frightens these unclean creatures into fearing all such forces of that region, as they believe that from that location their punishment is either being prepared, or will come."*

The mention of the lodestone, or magnet, would seem to indicate an electromagnetic connection. Another interesting fact is that iron is almost never mentioned as being a material seen in use by entities or in ships in UFO cases, whereas magnesium, aluminum, and other metals or alloys are, some of which may be mistaken by "abductees" for "stainless steel," due to color and sheen. Other modern researchers in this area have indicated that the "reptoids," or underworld coalition of reptilian and gray aliens, have a fascination with, and desire to own, all of the powerful, naturally-occurring magnets that can be found. Is this due to the natural ability of these objects to disrupt various fields which are generated by a prehuman technology? This may not be effective against all subterrestrial denizens, however, as will be seen later.

What happens to the abducted when they no longer prove genetically useful? Illustration by: H.J. Ford

Also of note is the mention that the underworlders are in fear of another subterranean region in the far north, "as they believe that from that location their punishment is either being prepared, or will come at some future time." Both the Norse underworld of Nifleheim, and the Finnish Tuonela or "Dead Land," were believed to be located in a vortex or void somewhere in the icy reaches of the ArcticSea. In the biblical book of Job, we find:

Job 26: 7-8:

He stretcheth out the north over the empty place ("void" in the Revised Standard Edition), and hangeth the earth upon nothing. He bindeth up the waters in his thick clouds; and the cloud is not rent under them.

Job 26:10:

He has described a circle upon the face of the waters, at the boundary between light and darkness. (Revised Standard Edition)

and

Job 38: 17:

Have the gates of death been opened unto thee? or hast thou seen the doors of the shadow of death (translated "gates of deep darkness," in the Revised Standard Edition)?

In the Apocryphal Book of Enoch, Chapter XL, Verse 9, we find:

And I measured out the whole earth, its mountains, and all hills, fields, trees, stones, rivers, all existing things I wrote down, the height from earth to the seventh heaven, and downwards to the very lowest hell, and the judgment-place, and the very great, open and weeping hell.

The Book of Enoch relates the plight of the one who has been cast out and down, in Chapter XXIX, verse 4:

And I threw him out from the height with his angels, and he was flying in the air continually above the bottomless.

In Chapter IX, verse 1, is the approximate geographic location as well as a description:

"And those two men (angels) led me up on to the Northern side (of the planet?), and showed me there a very terrible place, and there were all manner of tortures in that place: cruel darkness and unillumined gloom, and there is no light there, but murky fire constantly flameth aloft, and there is a fiery river coming forth, and that whole place is everywhere fire, and everywhere there is frost and ice, thirst and shivering, while the bonds are very cruel, and the angels fearful and merciless, bearing angry weapons, merciless torture, and I said:

"Woe, woe, how very terrible is this place."

Is this the place that the "faeries," the sidhe, and the "dragons" fear, as their eventual destination? The New Testament book of Revelation (Apocalypse) 5:13 leaves no doubt from the Judeo-Christian point of view that subterranean races exist beneath both land and sea:

> *"And every creature (race or people) which is in heaven, and on the earth, and under the earth, and such as are in the sea,...."*

Other New Testament verses leave no doubt as to the existence of a hidden race or races, or species:

> *"No man in heaven, in earth, neither under the earth was able to open the book, nor look thereon." Rev. 5:13.*

> *"That at the name of Jesus every knee shall bow, of things in heaven, and things in earth, and things under the earth;" --Philippians 2:10.*

This latter verse was verified by Reverend Kirk, who claimed to have extensively observed the habits of the fairies. He said:

> *"They are believed to have nobles and a system of law, but do not display visible religious worship, love, or dedication to God the Blessed Creator of all things. They vanish when His name is called upon, or the name of Jesus is invoked, as all among them that dwell above or inside the Earth bow either willingly or by force (as in Phillipians 2, verse 10), and are rendered powerless at the instant they hear the Holy Name."*

Further elaboration on and verification of the observed electromagneticconnection comes from "Superstitions of the Highlands and Islands of Scotland" by John Gregorson Campbell (1900), from the topic under the heading "Protection against Fairies: "

> *"The great protection against the Elfin race (and this is perhaps the most noticeable point in the whole superstition) is Iron, or preferably steel (Cruaidh). The metal in any form--a sword, a knife, a pair of scissors, a needle, a nail, a ring, a bar, a piece of reaping-hook, a gunbarrel, a fish-hook (and tales will be given illustrative of all these)--is all-powerful. On entering a Fairy dwelling, a piece of steel, a knife, needle, or fish-hook, stuck in the door, takes from the Elves the power of*

closing it till the intruder comes out again. A knife stuck in a deer carried home at night keeps them from laying their weight on the animal. A knife or nail in one's pocket prevents his being `lifted' at night. Nails in the front bench of the bed keep Elves from women `in the straw', and their babes. As additional safe-guards, the smoothing iron should be put below the bed, and the reaping-hook in the window. A nail in the carcass of a bull that fell over a rock was believed to preserve its flesh from them. Playing the Jew's harp (tromb) kept the Elfin women at a distance from the hunter, because the tongue of the instrument is of steel. So also a shoemaker's awl in the door-post of his bothy kept a Glaistig from entering."

To return to Japan, whether due to needle-metal or not, Mimoto never saw her dragon-lover again but she did give birth to a hybrid child, whom she called Akagire Taro, or Chapped Son. This was due to the fact that his skin was cracked, creased, and scaled like that of a reptile. From here the ancient tale enters historic accounts, for a direct descendant of Akagire Yataro, as the son was known in manhood, was a member of the Genji Clan named Saburo Ogata, who took pride in the fact that he had scales on his body as had his ancestors before him. He was the grandson of Yataro the Fifth.

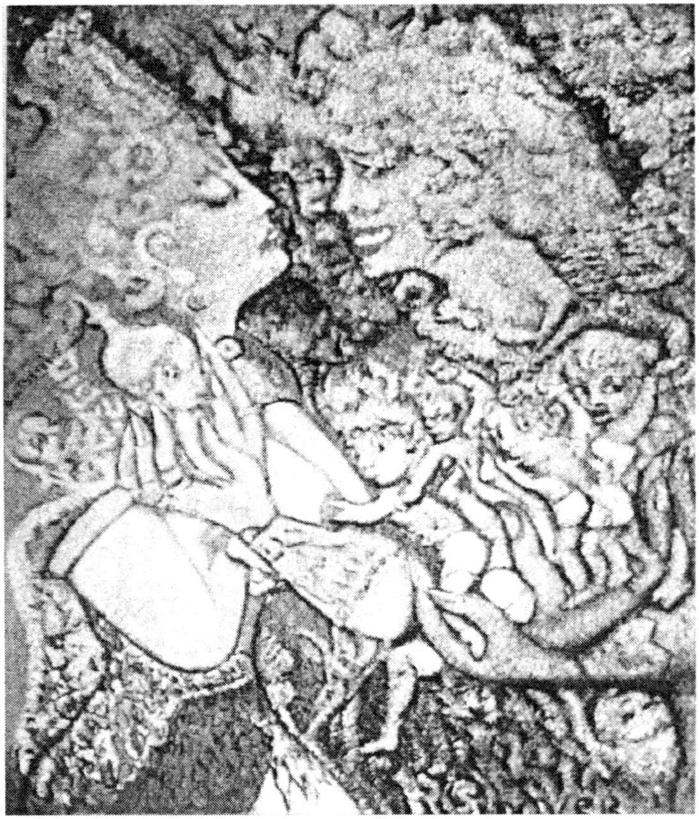

Shaver sometimess projected light through wafer-thin slices of his 'rocks,' using a projector. He would then create images on chemically-treated canvas of things he believed he had found, often in slices of rock of miniscule size, and then he added color.

In this picture, two of his titan-like 'Atlans' are shown, filled with joy over their genetic prosperity and bounty. This fascination with procreation and genetic diversity / success is by now well-established. According to Shaver the descendants of these 'Atlans' and 'Titans' wen underground (or were they driven there, as were the Nephilim of old?) in order to escape cosmic and solar radiation. Over the ages they degenerated into a variety of stunted and mis-shapen forms.

(See Color Inserts)

(See Color Inserts)

The Dragon-King pursuing the lady pearl-diver Tamatori. He has abandoned his human form in his rage/lust.
Illustration by: Utagawa Kuniyosha

Again, a prominent family line seems to have been the desired target of the original, and perhaps repeated, genetic exchange. While in the East, the influx of "dragon-blood" is seen as a thing of great pride, in the West such things are covered with an elaborate coating of fable and mist, becoming "fairytales" about serpent or frog-princes (as will be discussed later). Western sentiment, at least on the surface, is against such liaisons, often for religious reasons but not always on this basis alone, as the subterraneans have a track-record of cruelty, selfishness, and malice. While the Japanese "dragon" or reptilian humanoid introduced his genes into the primate pool, the mythical medieval knight Tannhauser was cursed by the Pope for his dalliance with a subterranean maiden, and when God forgave him by virtue of a Sign, he could not be found--he had returned in despair to the "Mountain of Venus," filled with underworld lovers who were his at the price of his eternal salvation, as he was doomed by a callous religious leader to become sperm-donor to a subterranean race.

This parsitical relationship has very ancient roots indeed. The most ancient Hebrew traditions center around the female demoness named Lilith, from the Sumerian Lil and the Babylonian Lilitu. She was the original "fatal attraction," as Hebrew esoteric tradition states that before Eve was created, Adam took a beautiful humanoid woman (not human, as she was not descended from Adam and Eve--perhaps neanderthal or another proto-human form) as lover or wife. According to various traditions and tales, Lilith was not the domestic type; she rebelled against the dominion of Adam and God, and after eating her own infant, she fled to a cave. There she took up cohabitation with the demonic residents who were already present, and began to breed with them. The primary bone of contention for Lilith was being subject to the rulership or authority of Adam, and her habits were decidedly feminist and darkly matriarchal in nature. One of her final promises or threats to Adam and Eve was that she and her offspring would always emerge from the underworld and wastelands, to abduct human children for their own purposes. Could this be an ancient record of a real and ongoing predacious activity? Perhaps the offspring of Lilith, a hybrid mix from the beginning, desperately need a periodic infusion of human genetic material or otherwise they become less and less "human" in appearance.

Michael banishing the devil – reptilian forces – Beneath the earth and the waters. Symbolic of primate/human ascendancy? Illustration by: Gustave Dore

It's interesting to note that Lilith is considered in esoteric tradition to be the ancestor or mother of the succubi and incubi, who of course engage in procreative or sexual activities with human beings. And is this all somehow related to "later" events, involving an upright, bipedal serpent and an piece of fruit?

It should be noted that the "serpent--" Satan, Lucifer-- did not "go on his belly in the dust" until after the events which led to the Fall of Mankind, but walked upright, coming and going in humanoid, bipedal fashion. The Torah and the New Testament both refer specifically to Satan as "that old dragon, the devil," "the serpent," and so forth. This begs the obvious question of a prehuman yet terrestrial race or form, of such antiquity as to have been originally of theropod, or bipedal saurian, derivation. Satan ("The Adversary") was lord over the Nephilim (literally "Those Who Fell From the Heavens,") who were generally regarded as fallen angels/demons and the losers of a cosmic war, and who were also said to be "in the earth."

Was this the race that offered refuge to Lilith, ancient yet terrestrially-bound refugees and rebels from a galactic conflict? The phonetic similarity between "Nephilim" and "Niflheim," the reptilian-monster-haunted underworld of the dead from Norse mythology, is interesting to note. According to Hebrew tradition, Satan was lord of the Earth in the distant past, before his rebellion; and as part of the punishment he received for his assault on "the Third Heaven (a greater galactic kingdom?)," he was "thrown downward to the Earth" where all that he had ruled over was destroyed. Now Science tells us that a sudden and catastrophic blow, perhaps an asteroid slamming (having been directed?) into the Earth, wiped out all of the theropod, brachiopod, saurian, and most other reptilian forms that existed in "the age of the dinosaurs." This theory is supported by an amazing new book, written by an avowed religious agnostic, who has been a reporter for the Washington Post as well as for the Wall Street Journal. Reporter Michael Drosnin spent five years studying the findings of Israeli computer scientists and cryptographers, who claim to have discovered hidden messages encoded into the Hebrew text of the Torah, or Old Testament. These messages are often in the form of the names of persons, places, or historical events which have occurred over the past few thousand years (and much earlier, it seems), as well as being of decidedly non-Hebrew origin or derivation. Both Drosnin, an eternal skeptic, and his Israeli contacts were no doubt amazed at what they found. To quote from page 148 of his book, The Bible Code, Drosnin writes:

"'Asteroid" and "dinosaur" are encoded together in the Bible. The Biblical name for the first creatures God created on Earth is also encoded in the same place."

"'And God created the great Tanin," states the (encoded) message (in) the first chapter of Genesis. The word means "dragons" or "monsters." It describes some huge animal that no longer exists."

"And "dragon" is encoded across "dinosaur," just above "asteroid." With them is encoded the name of the dragon that, according to legend, God slayed (sic; slew) before Creation."

"It is surely intentional that the name of the dragon the Bible says God fought--"Rahab--" appears in the Bible code exactly where the "asteroid" hits (crosses) the "dinosaur."

"In fact, the full hidden text states, "It will strike Rahab."

"It suggests that the extinction of the dinosaurs was the real slaying of the dragon (Fall of Satan and his kingdom at the time--W.M.M.), the cosmic event recalled by Isaiah: "Was it not you who cut Rahab in pieces, and pierced through the dragon?"

Drosnin's book also includes the actual Hebrew text, with the encoding revealed as it actually appears in the Torah. The implication is obvious: The "Fall of Satan" was synonymous with the "Fall" or destruction of the dinosaurs, and the end of the age of reptilian dominance on our planet. Rather than being a name for God's adversary, "Rahab" may have been the name by which the Earth was known before this event, or the name of the prehuman (saurian) yet humanoid civilization which existed then.

Since that time, the "reptilians" and their various offshoots have been in hiding, or banished beneath the Earth:

And the great dragon was cast out, that old serpent, called the Devil, and Satan, which deceiveth the whole world; he was cast out into the earth, and his angels were cast out with him. --Rev. 12:9

The Devil, both mammalian and reptilian.
The child – genetic bounty-- is the focus of attention here.

Satan may be a being or entity in his own right, but his name is also synonymous with the reptilian races he had lorded over (but who were still created as part of the "Tanin" or reptilian kingdom, by God), and who joined him in his galactic, interplanetary, and spiritual rebellion. Lilith may or may not have been a descendant of this race or species, but those who took her in most certainly were. In the 50 million years which have passed since the destruction of the "Kingdom of Rahab," the subterranean descendants of the saurian humanoids have evolved or taken on a variety of forms, racial types, or species, some apparently as comely and humanoid as "fairy princesses," others as hideous as hairy humanoids, trolls, and enigmatic forms like "grays," "reptoids," and "El Chupacabras."

Satan was also called "The Prince of the Powers of the Air," and was regarded as ruler over a wide range of kingdoms above and below the earth, all of which had vehicles for traversing the underworld, the land, the sea, and particularly the firmament. In Zechariah, Chapter 5, we find the prophet speaking with an angel of the Lord:

> *"Again, I lifted up mine eyes, and looked, and beheld a flying scroll (cylindrical or "cigar-shaped" UFO?). And then he said unto me, What seest thou? And I answered, I see a flying scroll; the length thereof is twenty cubits, and the breadth thereof ten cubits. Then he said unto me, This is the curse that goeth forth over the face of the whole earth: for everyone that stealeth shall be cut off as on this side according to it, and everyone that sweareth falsely shall be cut off as on that side according to it."*

Such intrusive sexual predators and meddlers in human affairs, seemingly paranormal in nature, are almost invariably associated with mysterious lights, wisps, and fires--Zeus and his lightning, the dragon and its flames or nimbus, the faeries with their will-o'-the-wisps and "witchlights." The parallel with modern UFO sightings of various sizes and types is clearly obvious, and perhaps indicates an effect which is more visible by night.

It would be comfortable and oh-so-smugly complacent and superior of intellect to look upon these accounts, and thousands of others like them from folklore and fable from around the world, and dismiss them as the primitive imaginings and musings of more primitive minds. The only problem with such a position is the fact that, according to many people living in the here and now, such things are continuing to take place on a regular basis. Not only is there a direct connection between the UFO phenomenon and some subterranean myths or reality, there is apparently a continued fascination with genetics, abduction, and offspring.

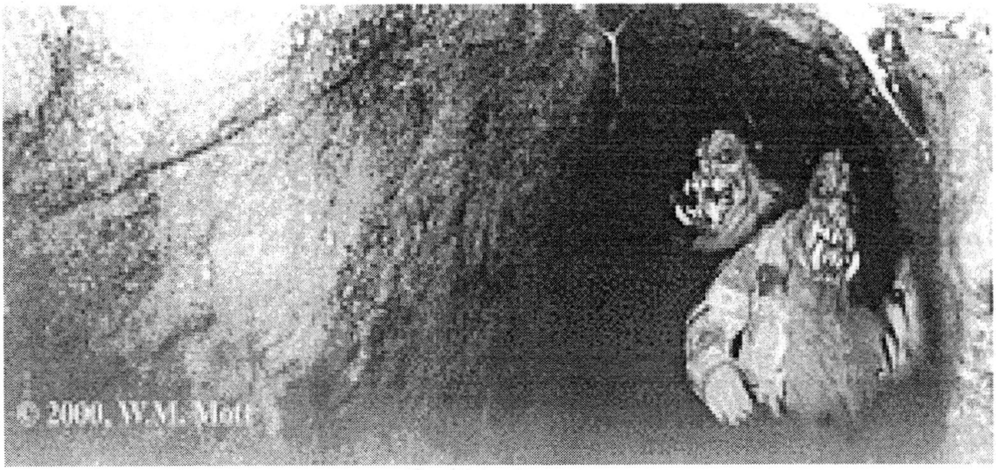

(See Color Inserts)

2. THE MORE THINGS CHANGE, THE MORE THEY STAY THE SAME

Norman L. (his name changed at his request), a single father living in the mountainous region of North Alabama, is one person who can testify to the truth of this continued intrusiveness. His wife had become deeply involved in caving (spelunking), leaving Norman and their toddler daughter to care for themselves most of the time. The wife became engrossed and fascinated with the past-time of caving, and often reported to her husband of seeing UFOs during some of the expeditions made with her hard-partying friends, multiple-day trips into remote and sometimes almost-inaccessible areas near "wild" caves and pits. The marriage began to deteriorate for a variety of reasons, many of them traceable to the obsession of the wife for spelunking; and strange phenomena began to occur around the house when the wife was home, or on the rare outing with the family.

Around nine one evening this family of three paid a visit to a local department store, one of the type which incorporates "mart" into its name. This store was located near a large ridge of mountains which ran alongside the eastern edge of the town, mountains which are honeycombed with caves and pits. The oldest part of the town itself is said to sit atop an ancient network of caverns, around which many local legends revolve. So husband, wife, and child cruised the aisles of the supermarket, when both adults became aware of an intense sense of presence.

The back corner of the store was unusually dark, as if some of the lighting were either dimmed or out. At the end of an intersecting aisle a strange figure lurked, literally leering at them as they walked by. According to Norman L., this figure was hideously malformed, in that he had no discernable neck, almost no chin, misshapen ears and facial features, and a decidely evil, grimacing disposition. His eyes were squinted and piggish, his head literally "squarish." His rounded, hunched body was wrapped in very loose clothing, primarily an immense flannel shirt which hid the true contours of his body. His attitude was definitely menacing as he transferred his gaze from the man, to the child riding in the shopping cart. He leaned forward as if to get a better look.

Norman stepped forward defensively, even though he was filled with an instinctive revulsion by this being. He insists that the person they encountered wasn't human, an opinion that his wife at the time had concurred with. Norman's take on this strange voyeur?

> *"He simply wasn't a human being. He looked human in a fundamental, first-glance kinda way, but he was something else. He had a strong air of evil and hate about him. He actually looked like a troll."*

Upon seeing Norman's reaction, the creature slunk back into the shadows of the aisle and was soon left behind.

Over the next year, the marriage dissolved and ended. Norman ended up with custody of the daughter, and took her with him everywhere. One night they visited a below-ground-level toystore in a local multi-level shopping mall; with the exception of the sales clerk standing near the front of the store, they were the only people in the place. Or so it seemed.

As his young daughter squatted to play with toys on the bottom of the shelf, Norman became aware of another presence. He turned to see a small child slowing meandering down the aisle toward them, seemingly a thin little girl of about five or six years of age. Suddenly this "child" made a quick move toward his daughter, and Norman scooped her up and took several steps back.

**The king snake seems to prefer the company of young women, in most cases.
Illustration: H.J. Ford**

> *"It wasn't a little girl," he says. "It was wearing a dingy yellow dress, had a girl's shoes and socks on, and its hair looked loose on its head, like a straight black wig, very artificial in appearance--kind of like a "Prince Valiant (haircut)." Its skin was abnormally pale, the eyes wide-set and almost at the side of its head, 'wrapping around the edge,' sort of. The eyes were almond-shaped, turning up at the outer edges, but squinty, and they were pure black or dark brown. There were no whites to the eyes at all! They were like ink."*

Thinking that perhaps this being was an unsupervised, malformed or mentally-handicapped child, Norman took his daughter and moved over an aisle; the little being followed them almost immediately. This was repeated twice, but the being stood off, a bit warier than before, still staring fixedly at Norman's daughter. He picked her up and they looked around the store for the parent of the "disturbed child," but there was no other shopper in the place. Aggravated, Norman complained to the cashier at the front of the store, asking that he have the parents of the child come supervise "it."

The cashier reacted with complete surprise. "You and your daughter are the only customers in the store, sir," he said. A quick look determined that this was in fact so, and no one had exited by way of the only way of egress in the front of the business. A quick look through the store by both men failed to reveal the strange "child."

Over time, many additional facts came out, largely gleaned from the daughter, who had been to visit her mother upon occasion. The child was terrified to make these visits, indicating that, among other things, she had been taken into wild, dangerous caves, leaned over pits (and allegedly almost dropped into one, to luckily be caught by the mother's boyfriend), and other such insanity. Eventually the mother's visitations were terminated completely, but this raises the question of motivation and influence--what would lead a mother to risk the life of her child in such a callous, almost sacrificial way? The child also witnessed UFO manifestations during one of the caving trips with her mother. Since that time, the girl has grown to be a happy and healthy teen, and no further instances of this type of encounter have troubled her, or her father.

First-hand accounts like this abound. Another account, also from the mountains of east-central Alabama, involves a group of children (now adults) who saw what they termed "a leprechaun," following them through the woods one evening. Initially noticed as he sat atop an outcropping of mossy rock, this entity hopped down from his perch as the kids passed, maintaining a respectful distance, but he paralleled the children for miles, scaring them badly. He was described by the now-grown woman and two men as being some two or three feet in height, wearing green clothing, and looking quite sinister.

It's interesting to note that the TAG (a National Speleological Society term) or Tennessee-Alabama-Georgia region is considered to be one of the most densely caverned areas in North America. Many accounts of mysterious lights, UFOs, livestock mutilations, and strange deaths and disappearances have come from this region. Tales of "haints" and "boogers" abound in the hills and mountains, and some spelunkers joke to cover up their fear when they talk about "hodags," unseen cave entities which cut rope or pull it up behind cavers after a descent; drop or throw rocks and boulders; mislead with mysterious noises; and create other such strange occurrences.

A similar account comes from Oklahoma City, OK. One lady indicates that she and her child are the objects of unwanted desire or attention from subterranean beings. She is extremely lucid and utterly convinced of the reality of the experiences she and her child have had.

The entire sequence of events started when she was camping with her four-year old daughter in a remote area. The daughter mysteriously disappeared, to be eventually found a few days later, "well-fed, clean, and happy." The little girl claimed that she had been the guest of an "underground family" who had been very nice to her and had fed her.

Does tradition not indicate the danger of eating "fairy food?" To do so is to become possibly "trapped in fairyland...."

In the ensuing three years since this mysterious event, the child insists that these subterranean people come to her at night as she dreams, and urge her to come outside, where they are waiting for her. They've indicated that they want her to go with them, back beneath the ground. The mother indicates that strange phenomena and experiences have been plaguing her and her daughter, and although they've moved repeatedly, the subterraneans always find them right away and resume the harassment. The lady is considering moving out of the state, which may or may not alleviate her problem; even more dangerously, she has sought refuge in some forms of mysticism and magic, which is of course traditionally considered the realm of expertise of those from "below." This was known to many ancient peoples as well, particularly the Hebrews:

In Psalm 74:20, we find:

> *"Have respect unto the covenant, for the dark places of the earth are full of the habitations of cruelty."*

The "covenant" would have been the Hebrew's particular covenant with their God, including the various prohibitions against divination, witchcraft, and so forth. Robert Kirk addressed the consequences of this in The Secret Commonwealth, averring that to assume the mantle of clairvoyance or second-sight is to abandon the protection of God, and to be under the power of the beings which are spied upon:

> *"The men with Second Sight (clairvoyance) discern mysteries when queried, by going into a trance-like state, as if possessed at that moment by some familiar spirit. I have often spoken with one of these who, in his trance-state revealed that he had cut the form of one of these beings in twain with his iron blade, and thus escaped the trance (or channeling state). But when he came out of it he saw no trace of the split body he had thought to see previously (while in the beginning of the "trance"). Upon other occasions, he out-grappled others of their kind when they seized him."*
>
> *" His neighbors regularly watched this fellow vanish at a specific location, and then reappear around an hour later, finding himself almost a bow's shot distant from his vanishing-point. It was in the place that he became unseen, he explained, that the subterrestrials did meet and fight (abduct) him.""*

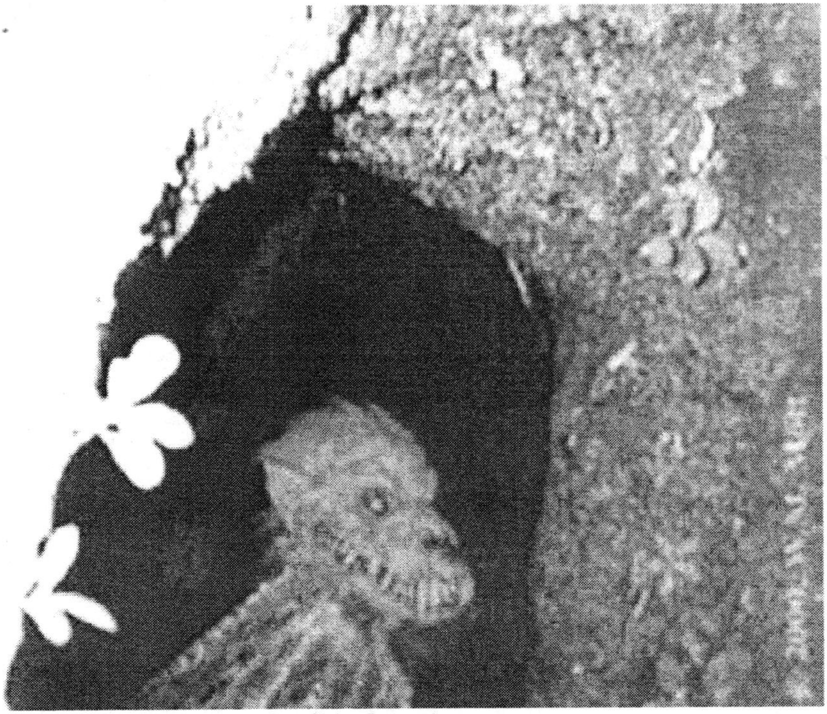

(See Color Inserts)

J.G. Campbell relates a tale similar to that of the lost Oklahoman child:

> *"The Elves sometimes took care of neglected children. The herd who tendered the Baile-phuill cattle on Heynist Hill sat down one day on a green eminence (cnoc) in the hill, which had the reputation of being tenanted by the Fairies. His son, a young child, was along with him. He fell asleep, and when he awoke the child was away. He roused himself, and vowed aloud, that unless his boy was restored he would not leave a stone or clod of the hillock together. A voice from underground answered that the child was safe at home with its mother, and they (the `people') had taken it lest it come to harm with the cold."*

Campbell is not clear on whether or not the child was returned to its parents, but if it was, it was probably harassed repeatedly as in the case of the mother and daughter from Oklahoma City. One aspect of both "fairy abduction" and "ufo abduction" " is that once the little buggers get a taste, they seem to keep coming back for more.

The occurence of these "little people," subterraneans, and so forth, goes back into antiquity all over the Earth, and particularly in the British Isles. Regarding the Scottish Picts, a very real pre- Celtic or partially-Celtic people, one Icelandic scribe writing in the 12th century chronicle Historia Norvegiae, stated that the Picts were a race of pygmies, who were known to "do wonders in the mornings and the evenings, but at mid-day lost their strength and hid in holes in the ground."

The actual tribes called "Picts" were not pygmies, nor did it seem that any of the ancient human beings of Scotland were particularly laid low by the noonday sun, so this Norseman may have been talking about some other group or race of beings, similar to the huldre or hidden folk of his own traditions. The increased activity in the pre-dawn and twilight hours, the hiding from the sun in holes in the ground, would all seem to tie these so-called "Picts" in with the Sidhe (shee), the gnomes, goblins, faeries, and so forth. Even Mimoto's dragon-lover preferred to come out into the surface world by night. Ireland and Scotland in particular were haunted by communities of these beings, as Kirk also stated:

> *" Even in this modern time (Kirk's time of writing) they are at times noticed to be baking bread, striking with hammers, and undertaking similar domestic activities in the small hills which they frequent most often. They are of an ancient epoch of time before the advent of Christianity and the end of paganism, and in some wild barbaric places still, they come into houses when the inhabitants are asleep, cleaning the kitchen and all of the dishes. This variety are called Braunies (Brownies). When we humans have abundance, they have little in their abodes; and the inverse is often true, although they are not allowed a free rein in their predations. Their thefts aside, they can many times be seen interfering with the fertility of the grains and crops, influencing either scarcity or overabundance."*
>
> *The Icelandic Chroniclers were no strangers to the belief in underworld beings, being of Norse extraction themselves. As an island-nation with isolated genetics, they too seemed to be particularly plagued*

by underworld or faerie activity. The belief that some strange race in Scotland might live "in holes in the ground" would not have been excessively exotic, nor would the tales of aversion to sunlight. The Icelandic sagas are filled with incidents of interference and conflict with hudre (hidden) or elf-folk, and trolls. An account from Grettir's Saga (Icelandic, 14th Century) indicates how familiar the Viking peoples were with the habits of the subterraneans, and their matter-of-fact attitude about them. In the following passage (Translation from the icelandic by G. H. Hight, London, 1914), Grettir has killed a man during the daytime, and initially blames the death on a female troll or ogress:

"Soon Skeggi was missed in the company, and when Grettir came up they asked him what news he had of Skeggi. He answered in a verse:

Hammer-troll ogress has done him to death.

Thirsting for blood the war-fiend came.

With hard-edged blade she gaped, o'er his head,

nor spared she his teeth. I saw it myself.

"Then Thorkell's men sprang up and said it was impossible that a troll should have taken the man in full daylight."

Apparently they were quite familiar with the habits of trolls. An excellent example of these habits is described in the tale of Peer Gynt, as immortalized by Henrik Ibsen in his play of the same name; but the story is based on an earlier Norwegian tale or series of similar tales, which in turn are based on the Scandinavian troll and huldre traditions. Peer, the ne'er-do-well son of a hardworking mother, forsakes a good and loving betrothed for dalliance with troll-maidens, and eventually a tryst with the green-clad daughter of the Dovre(troll) -King himself. A night of debauchery leads to nearly being ensnared in an unnatural marriage, and eternal entrapment in the underworld kingdom of the troll-king. No better than the trolls in some respects, Peer does what he is best at doing--he breaks his word--and manages to escape a host of subterranean enemies, as well as a horrible, captive fate as stallion to the troll-woman, thanks to the sound of church-bells and to distant prayers said for his sake. Later in the play or story, however, he meets the troll-woman in her true form and does not at first recognize her; and she has with her their half-human son, who has one malformed leg. The troll-folk under the hill also have tails and other non-human characteristics. Deformities or non-human irregularities of the extremities are a regular feature of the subterranean folk, as will be demonstrated further.

The serpent, the babes and their genetic potential, the tree of knowledge (reproductive), and the warrior/father figure.
Illustration: H.J. Ford

Another contemporary account which has come to light is from Timothy H., who lives in Idaho, but is partially of Scottish descent. As a child he went to visit his grandfather in Scotland, where he had a most unusual encounter which would seem to indicate that there may be two or more factions of such beings (or perhaps numerous nations), some malevolent, some more-or-less friendly, apparently as the mood hits them or as it suits their purposes.

While riding his tricycle along a pathway near Inveraray, he took a hard tumble, and thinks to this day that he probably broke his arm. After he'd been sitting there alone and crying for a few minutes, a little person--no more than two and a half feet tall--appeared out of nowhere to assist him. Somehow he tended to Tim and healed his injury almost instantly. Tim says he asked the little man who he was, and the reply came that he was a gnome; the boy observed that he was a leprechaun, and the little man corrected him patiently, saying that no, he was a gnome. About this time Tim's parents, who had been looking for him, came near to the scene, and the little man scurried away into the underbrush. The parents didn't believe their son's story, but his old Scottish grandfather did, and he told Tim that only people of Gaelic descent, and most often children, see the subterraneans. This may be true in Scotland, but tell that to the people in Alabama and Oklahoma! But again a bloodline-tie or interest is indicated by this story, wherein which members of a particular ethnic or tribal group are singled out for attention from unknown forces.

Obviously, not all subterraneans are diminutive. Often they more closely resemble Mimoto's lover. Another lady from Oklahoma says that as a child, forty or more years ago, she was 'befriended" by "very tall people" who lived in an underground river tunnel which exists beneath Oklahoma City. She would play in the water at the edge of the tunnel entrance, and they would "watch over her" as she played, "keeping snakes away." Apparently big snakes scare away little ones, because she also says that, like the Ryu in the Japanese story, these beings were "not human-looking."

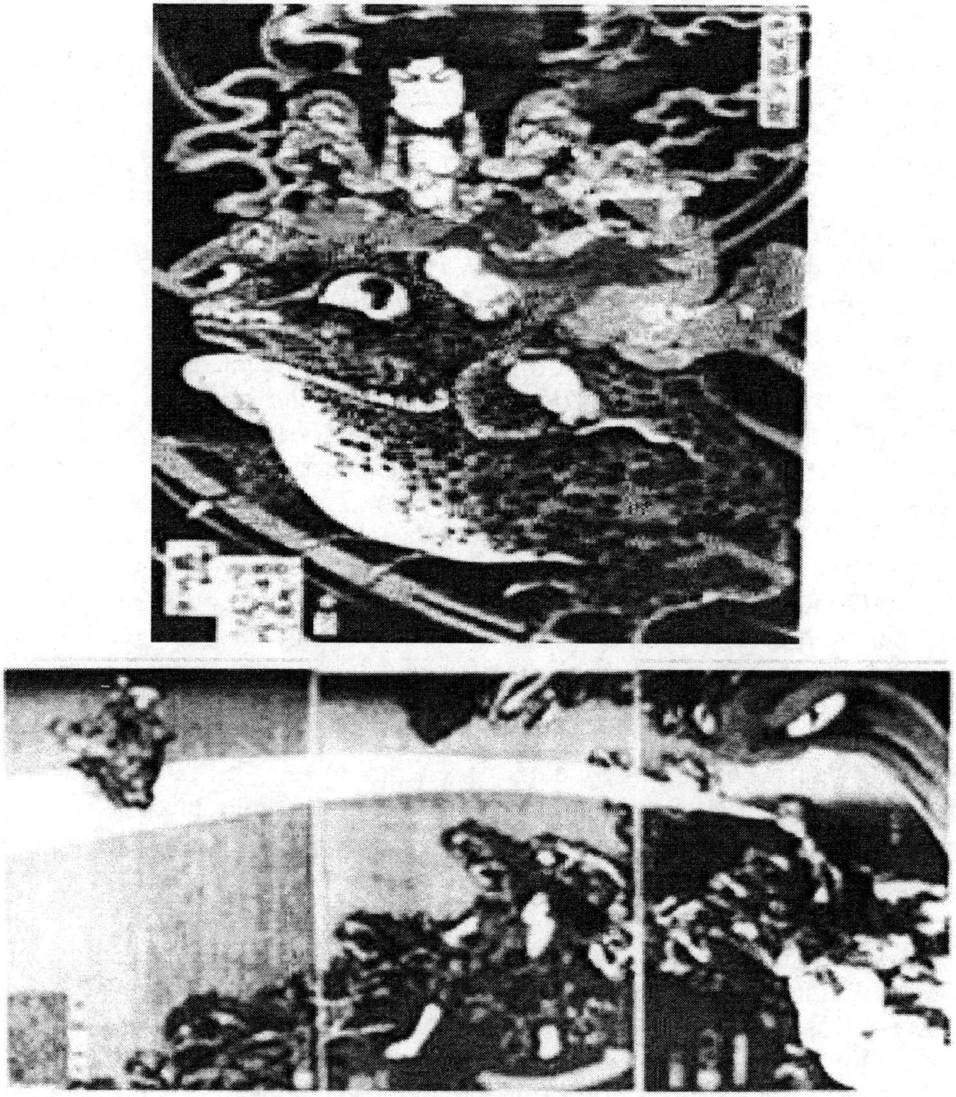

Japenese legends are filled with tales of shapeshifting sorcerers who take the form of giant toads. 'Toad magic' is associated with the underworld, and is considered to be one of the most evil types of sorcery

The reptilian aspect of some underworlders permeates folklore. One universal theme that recurs in the folktales of many, many cultures is that of the snake-husband or snake-wife, who can transform into a "human" or humanoid form and is invariably (of course) of royal blood among his or her own kind (talk about the ultimate pick-up line!). Often the snake or serpent-man exacts a promise of marriage, or the hand of an unborn human child in betrothal, consistent with the theme of the subterraneans' interest in maintaining their own genetic diversity. A variant of this should be familiar to most readers of fairytales, in the form of "The Frog Prince." The frog-prince is a Handsome Prince, but like the Japanese seducing dragon, he has a reptilian or amphibian form. The underworld link is complete, for frequently the frog lives in a deep well, from which he is discovered or rescued by the female protagonist. A possible connection is evident in the Scandinavian belief that some dwarves would "turn into toads," if caught by the sun, much like Mimoto's lover turned from a man into a "dragon" when the same thing happened. Slovenia has its legends of fairies and "little people," but Slovenian fairytales are also permeated by the presence of the "Snake Queen," a great, white, cave-dwelling creature who is part woman and part serpent. The serpentine-yet-human Nagas are still believed by devout Hindus and some Buddhists to dwell beneath India, Nepal, and Tibet. And of course, if we consider the connection between the fairytales of recent centuries, and the ancient knowledge of Lilith, the Nephilim, and Rahab, it would seem that perhaps St. Patrick missed a few snakes when he was cleaning them "out of" the hills of the Emerald Isle, after all.

Other reptilian connections exist, possibly too many to document in one volume. Some tales of nocturnal Germanic dwarves indicate that they hid a "deformity" from prying human eyes, that deformity being "goose-like" feet--that is, their feet are webbed with four amphibian-like toes (Wilhelm and Jacob Grimm, German Legends). This is a common motif which differentiates "supernatural," faerie, or other strange beings from humanity, particularly in European folklore ("deformities" or "defects of the fairies," as Katherine Briggs calls them)--the Bean Sidhe (Banshee) has one webbed foot (as well as other deformities), the Scottish Glaistig hides goat's hooves beneath her long dress, the Shetland Henkies limp strangely when they dance, and so forth. Underworld imps and devils were said to have hooves as well. Webbed fingers, tails, and other non-human characteristics are a sure give-away of underworld genetics. Even modern "gray aliens" are said to suffer from similar deformities of the extremities, and move in a "shuffling" gait as if uncertain of their balance, much like the "Henkies." Some of these deformities could well be adaptations to the challenges of living underground, where sure grips are needed when clambering and climbing, and banged extremities might occur more often.

3. HAMBURGERS FOR HELL, AND OTHER NASTY DISHES

Most people in the western world have probably heard the story of the Three Billy Goats Gruff. Three goats of varying size are harassed by a hideous and voracious troll which lives under a bridge. This story originated in the high, mountainous region of Scandinavia, where a rich folk-belief in trolls and other supernatural beings exists, but it is by no means a purely Norse motif; the theft of livestock is a recurring theme around the world, as is the "cross-breeding" of livestock owned by humans with creatures owned by subterranean races.

Another aspect of this is livestock mutilation, a form of harvesting which may be, among other things, in order to obtain cells for cloning or purposes of genetic manipulation. As gruesome as it might sound, the obtaining of protein--food-- is another definite possible motive behind both cattle and human abductions.

John Gregorson Campbell, in "Superstitions of the Highlands and Islands of Scotland," relates many such examples of similar behavior, which bear some startling similarities to modern day "cattle mutilations" that are attributed to UFO activity:

"The Charmed Hill (Beinn Shianta), from its height, greenness, or pointed summit, forms a conspicuous object on the Ardnamurchan coast, at the north entrance of the Sound of Mull. On `the shoulder' of this hill, were two hamlets, Sginid and Corryvulin, the lands attached to which, now forming part of a large sheep farm, were at one time occupied in common by a three tenants, one of whom was named Luran Black (Luran Mac-ille-dhui). One particular season a cow of Luran's was found unaccountably dead each morning. Suspicioun fell on the tenants of the Culver (an cuilibheir), a green knoll in Corryvulin, having the reputation of being

tenanted by the Fairies. Luran resolved to watch his cattle for a night, and ascertain the cause of his mysterious losses. Before long he saw the Culver opening, and a host of little people pouring out. They surrounded a gray cow (mart glas) belonging to him and drove it into the knoll. Not one busied himself in doing this more than Luran himself; he was, according to the Gaelic expression, `as one and as two' (mar a h-aon 's mar a dha) in his exertions. The cow was killed and skinned. An old Elf, a tailor sitting in the upper part of the brugh, with a needle in the right lappel of his coat, was forcibly caught hold of, stuffed into the cow's hide, and sewn up. He was then taken to the door and rolled down the slope. Festivities commenced, and whoever might be on the floor dancing, Luran was sure to be. He was `as one and as two' at the dance, as he had been at driving the cow. A number of gorgeous cups and dishes were put on the table, and Luran, resolving to make up for the loss of the gray cow, watched his opportunity and made off with one of the cups (corn). The Fairies observed him and started in pursuit."

Before modern cattle ranchers lament that they've never received payment in gold for the pilfering and mutilation of their cattle, they should be glad to not have taken anything in payment or restitution:

> *"Their persecution of Luran did not, however, cease. While on his way to Inveraray Castle, with his fairy cup, he was lifted mysteriously with his treasure out of the boat, in which he was taking his passage, and was never seen or heard after."*

Much like their modern counterparts the "UFOnauts," the sidhe often used an inscrutable power of flight as their means of locomotion, and their swift abduction of victims. Often the fairies sweep through a region in a storm or whirlwind, abducting as they go. The Highland Gaelic peoples believed in two distinct groups of kidnapping fairies: the sidhe who were encountered walking on the ground after sunset, even as early as twilight, and who could still befuddle and abduct the unwary mortal, and the "Sluagh Sidhe", a malevolent fairy host which traveled through the air by night, and was known to take mortals with them for the ride. These latter are synonymous with both the demonic "Unseelie Court" of Highland belief, which does the same thing, and with (incredible as it might seem) the modern-day "alien abductors," be they human-like in appearance, or gray or reptilian humanoids of uncanny look. It should be noted again that Satan was also called "The Prince of the Powers of Air." What is not at first apparent is that these "powers," while atmospheric, may have subterranean origins.

John D. Batten illustration for 'The Laidly Worm of Spindleston Heugh.' from ENGLISH FAIRY TALES, collected by Jospeph Jacobs (David Nutt 1890). The girl is transformed from a 'dragon' and into her true (?) form.

How many 'beautiful maidens' had to be 'rescued' or 'found' by non-humans of subterranean origin? All they need is a UFO and a long thin needle. Illustration by H.J. Ford from the Olive Fairy Book, by Andrew Land (Longmans, Green, 1907.)

Campbell and other well-known chroniclers of oral traditions recount numerous historically-recent accounts (as of their time of writing), some seen by witnesses, of fairy kidnappings of men, women, children, and animals. Sometimes the victims are returned, sometimes they are not. There is often an element of "missing time" experienced by those who are taken into the subterranean realm, time spent (as far as they can recall) in dancing, feasting, singing, or attending to fairy births or fairy children. Some of these "memories" may or may not be confabulation. The aspect of missing time is a condition that is familiar to researchers into UFOlogy and abductions, which also have a theme of the theft of genetics or exploitation of the human gene pool. The "fairy children" might just as well be called "human-alien hybrids," as the ancient and contemporary descriptions of both types of beings seem to be identical.

To return to the billy-goats, what are the additional meanings of the tale, if any? Bridges span chasms and gulfs, hide rock-faces, and perhaps caves or clefts, with shadows which keep off the sun. Jeremy R., now a worker at an internet-based company, had a curious story to relate which involved a bridge, taking place in 1984.

> *"I was driving from Birmingham, Alabama, to Jackson, Mississippi, when I had a blow-out on the interstate," he relates. "It was almost sundown, and the tall pines around, along with the hilly terrain, had created a dark atmosphere already. I was close to Meridian, Mississippi when it happened, and pulled over just past an exit and overpass. While I was working on changing the tire, I kept having an eerie feeling that I was being watched. You know, the hairs on my neck and arms were standing up (electromagnetic effect? --W.M.M.), my spine was tingling. Pretty soon this feeling became unbearable, and I jerked around to have a look, like (as if) by instinct. That's when I saw it, a short, stocky figure, about fifteen or twenty feet away, toward the overpass. He was wearing pants, I guess, and a big dark coat with a hood. I thought it was a kid, so I turned to speak to him, and I involuntarily raised the lugwrench--one of those cross- or plus-shaped ones, you know--and he started moaning. I mean, it was the creepiest sound I've ever heard! Then I saw his face, or got a brief look at it--pale, kind of reminding me of a fish or some other thing that didn't get much sun. Couldn't see his eyes, they were deep and sunken-in, shadowed. He gave me the creeps like you wouldn't believe, so I stood up, and he skedaddled out of there! I watched him run to the overpass, then scamper up the concrete embankment and disappear. I figured he was hiding out in the shadows. I thought he must be a midget, 'cause his face was shaped like a man's.*
>
> *"I'd just finished tightening the last lug-nut, so I hopped in my car and backed up the shoulder of the interstate to see what was going on. I carry a gun, a 9mm, so I had it out and in hand just in case, and backed up under the overpass, which was on more of a hill than where I'd been before, with the flat. The setting sun in front of me was shining up into*

the shadows, and for a second I saw him, or thought I did. He was hunched over under the pylons, and he dove back and just disappeared somehow! There was a break or pocket up there, no more than a foot or so in height, but I don't see how he could fit through it, he looked too bulky. I was freaked out, so I burned rubber outta there.

"When I headed home a few days later, I got off at that exit, since it was mid-day and the sun was shining, and I felt a little more brave. I parked at the top and walked down the off-ramp to check it out. I climbed up under the overpass, all the way to the top, right under the bridge, but there was nothing there! The hole was just an indentation, scraped-around and littered with leaves. It was only a foot or so deep, way too shallow for even the smallest person to jump into. I walked around, checking out the hill and soil under the grass, that the underpass was on, which was a hard, tough clay that I thought at first was rock. But there was no sign of a little man, a midget, a kid or anything else! But I know what I saw."

So what did Jeremy see? Was he hallucinating? He was two-thirds of the way to his destination at this point, and had probably been driving for at least two hours. If he wasn't hallucinating or having some sort of confused experience, then what is the explanation for the person's behavior, if it was a human being? Why did it not speak, why was it moving up on him stealthily? Why did it recoil from the IRON and the CROSS of his lug-wrench, just as a Scandinavian or Celtic fairy would have done, according to those traditions? How did it vanish, literally "into the earth?"

Old patterns do indeed repeat, it seems.

(See Color Inserts)

4. WHAT IN THE WORLD.....?

Let us now consider the Earth itself. We dwell upon the planetary skin like microbes on a wrecking-ball, or subatomic entities teeming on the surface of a billiard ball. Due to limitations of temperature (sometimes), pressure, and motivation, we've yet to more than vaguely explore the true depths of our own planet. We know more about the surface of Mars and Venus than we do about the bottom of our own deep ocean trenches, or about the true nature of the Earth's crust, which is apparently more often like a sponge or porous piece of bone than it is a solid substance. The thin layer of reality that we call the "surface world" is really a very limited domain.

(See Color Inserts)

The crust of the Earth, however, starting right below our feet and stretching down to the bottom of the Mohorovicic Discontinuity, is on average 70 kilometers, or around 43 miles in thickness. The Mohorovicic, or "Moho" as it's called, is an anomalous region of ambiguous density that upon seismic examination seems to be composed of hills and valleys, peaks and troughs--in other words, it is a vast, world-wide cavern system, probably varying in internal condition depending upon it's nearness to magma pockets, and lava-siphoning regions which reach down to the mantle.

From this point upward, the crust is absolutely honeycombed with concavities, cracks, fissures, and the like, from the breaks between tectonic plates and subduction zones, to aquifers both emptied and filled with water, oil and natural gas reservoirs, underground rivers and streams (and ancient riverbeds which have long-since run dry), and of course, caves and caverns.

Only upon occasion do we get a glimpse of this truly titanic realm, when cavern systems of awesome size are discovered. Mammoth Cave in Kentucky is one such massive system, and may be part of a larger system which underlies much of Eastern North America, particularly along the Appalachian mountain chain. Lechuguilla Cave, in New Mexico, is another. Central and South America have numerous cavern systems which have yet to be fully explored or plumbed, like Chiquibul Cave in Guatemala and Belize, believed by the Mayans to be the entrance to Xibalba, or the underworld. While recently explored to a great extent, Chiquibul is like other systems which obviously and undeniably link up via unknown tributaries or passageways with other, vaster, deeper systems.

Mammoth Cave is a limestone cavern system which formed in an ancient seabed that is now a mountain range. Mammoth has an extent of over 150 miles of explored passages, and it just goes on and on. Carlsbad Caverns, located five miles from Lechuguilla, has never been fully explored, and the full extent is not known. The guided tour route covers only three miles, and the system goes onward and downward into eternal darkness. By far the largest North American system yet explored, however, is Lechuguilla, a "blowing cave" of tremendous size. Expeditions go into it for days and weeks, yet the caverns seem to have no end. Like many large cavern systems, Lechuguilla is distinguished by a strong wind which blows with great force from it's entrance. What is the source of this wind, and why is it continuous? Blowing Cave in Arkansas is another example of this, as are many others. This seemingly-uninterrupted flow of tremendous volumes of air would indicate an ultimate source of both great pressure, and great volume. Do these cavern-systems reach down eventually to the Moho layer itself, where tremendous pressures exist?

Apparently there's a lot more going on beneath our feet than we tend to realize, and there always has been. Traditions around the Earth assert that there are other worlds locked away beneath the crust of our planet. The worlds of Yggdrasil, the Norse world-tree, upon close examination all tend to be subterranean in some regard. The Native American mythologies are filled with tales of origins and adventures in underground worlds, often multi-leveled, as are the traditions of Central Asia and the Far East. In Tibet, Nepal, and India, these realms are known as separate empires or kingdoms beneath the ground: Agharta, Shamballah or Shambhala (source for the fictional Shangri-La), Patala, Bhogavati, and others. Some are inhabited by humans, but most are populated by a wide variety of humanoids, gods, demons, reptilian beings, and hairy humanoids. The hirsute Yeti of the Himalayas are believed to dwell within the mountains themselves, in caves and gorges; the scaly Nagas dwell beneath the ground, and the entrances to their realms are often located beneath lakes and rivers. Native American traditions indicate that hairy humanoids dwell in caverns beneath the Earth, as do snake-people, ant-people, lizard-people, and the "moon-eyed people," a tall white race of the ancient eastern woodlands, which fled beneath the ground before the advance of the Indian ancestors.

Often people are shocked to find that the cities they live in are located above vast, and often largely unexplored, cavern-systems. The old downtown section of Huntsville, Alabama is one such place, as is Oklahoma City. Guthrie, Oklahoma, a small town of about eleven thousand people, sits atop a gigantic cavern system in which outlaws of the Old West were once accustomed to hiding. For some reason the city of Guthrie has sealed the entrances, and now denies that the caverns exist, but old-timers still know where the entrances are. Rumors of similar natural caverns, and artificial tunnel-systems, exist from Britain to Malta, and from Southern California to Wisconsin, and to New England. Lexington, Kentucky has its "catacombs," which were already there before the first Europeans arrived. Salt Lake City is another locale with rumored tunnels and caverns beneath, and the list goes ever on. And of course the wild or sparsely-populated places of the world are continuously found to contain new sinkholes, sumps, pits, and vortex-plagued lakes which apparently interconnect with one another, or with deeper aquifers and caverns.

Cryptid critters, UFO phenomena, and other creepy varmints seem to have a definite connection with the underworld, as do unusual human beings of "outlandish" or "foreign" appearance. The very nature of the rapid appearance and disappearance of such beings would indicate that not only caverns, but land-locked bodies or water, oceans, old cellars, mineshafts, train tunnels, bridges (over hidden water or land entrances?), sewer systems (particularly older ones) may have a definite connection to such things. The Toronto, Canada, Sunday Sun reported just one such incident in the March 25, 1979 issue.

(See Color Inserts)

Habetrot and the entrance to her cavern. From MORE ENGLISH FAIRYTALES, illustration by John D. Batten, Collected by Joseph Jacobs (David Nutt, 1894)

As the paper reports, and as reprinted in Shavertron magazine, a respected citizen of Cabbagetown known only as "Ernest" had a face-to-face encounter with one denizen of the underworld. Ernest didn't take his story to the paper or otherwise seek publicity; he was sought out by a reporter who heard about his experience from a mutual aquaintance. While searching for a lost kitten, Ernest clambered down into a cavern or tunnel beneath the streets near his Parliament Street apartment, through an entrance covered by an ancient slab of concrete; sliding the slab far enough to admit his full-grown human form, he suddenly entered a different realm. He related:

> *"It was pitch black in there....I saw it with my flashlight. The eyes were orange and red, slanted...it was long and thin, almost like a monkey....three feet long, large teeth, weighing maybe thirty pounds with slate-gray fur."*

As if this encounter with a strange "animal" weren't enough, the next stage of this bizarre meeting really took him aback.

> *"I'll never forget it," Ernest said. "It said 'Go away, go away," in a hissing voice. Then it took off down a long tunnel to the side, and I got out of there as fast as I could. I was shaking with fear."*

The reporter accompanied Ernest back to the scene of his encounter with the simian imp, or small hairy humanoid. The corpse of a full-grown cat lay half-buried in the tunnel, but the slab over the entrance had recently collapsed "during the winter," effectively sealing any entrance by a human being. Ernest believed the tunnel ran further downward, to meet the sewer system of the city, but this was neither verified nor denied by the city sewer department. One sewer worker stated that not only did he not want to go into the tunnel alone, but he didn't scoff at Ernest's tale.

As often happens in the case of the haunts of trolls, imps, and fairies of Europe, as well as in the instance of the bridge-haunter of Mississippi, the entrance was effectively sealed "by accident" or "erosion" after it was discovered by a human interloper. Doubtless the sewer department of Ernest's city finished the job of sealing it up, to prevent injury, or further weirdness! This type of "sealing" of natural or "unplanned" caverns and tunnels happens in urban areas more often than people realize.

Stories of encounters with subterrestrials stretch from antiquity to the present day. Glimpses into a vast subterranean biosphere with a great diversity among specie, largely protected or hidden from the surface-dwelling human race, have been recorded time and again. Apparently the "deep dwellers" of our planet vary widely in size, temperament, and form, some extremely human in appearance, others scaled, hair-covered, or otherwise horrific. In the 12 century, Gervase of Tilbury, a monastic English chronicler, recorded the report of two "children" suddenly appearing close to a village near Burry St. Edmunds, England. Gervase's tale was also recorded in the writings of chroniclers who were more less his contemporaries, such as Ralph of Coggeshall in the CHRONICON ANGLICARUM; William of Newbury, in the HISTORIA RERUM ANGLICARUM, and also the chroniclers Walsingham and Giraldus Cambrensis.

A Samuri fights a NUE – a mammal-reptile humanoid hybrid.
Illustration by Utagawa Kuniyoshi
(See Color Inserts)

The importance of these accounts should not be underestimated, as medieval Christian monks were not wont to outright lying or deceit, concerned as they were for their pious souls. Reports of strange happenings were more often than not investigated by the Church itself, before being categorized as natural, unnatural, or supernatural events.

The gist of the tale lies in the discovery of two disoriented, haggard young children, near the village of Wolf-pittes, who were found roaming ancient trenches or pits known locally as the "wolf pittes (pits)," for which the town was named. The children, a boy and girl, were hungry and exhausted, and did not speak English or any other recognizable (European) language. The characteristics which most surprised their rescuers and benefactors, however, were not their perfectly human form and faces--it was that their skin was an olive green color, and their hair was green as well.

An underworld suitor rebuffed.

Illustration: H.J. Ford

Despite their outlandish language and eerie coloration, the villagers took them in and they were sent to a local home to be cared for. At first, they refused all manner of sustenance, until someone brought in some beans, still in the pods; these they fell upon ravenously, and for a long time this was all they would eat. As Katherine Briggs points out (in AN

ENCYCLOPEDIA OF FAIRIES, 1976), their skin-color, green, is the color which the ancient Celts associated with death and the underworld; and beans were traditionally the food of the dead. This may or may not have been meaningful to their Norman-Saxon benefactors, however, but the connection with the ancient mound-folk traditions is interesting.

After some time, the "Green Children of Wolf-pittes," as they were called, took separate paths. The boy, melancholy since his rescue (or capture), languished and died, while the girl flourished, began eating more surface foodstuffs, and lost her green pigmentation. If of a hybrid (partially human/Adamic) stock, she and her brother may have had different levels of tolerance for surface-world environmental conditions. Whatever the case, she learned to speak English and lived to serve in the household of the knight, Sir Richard de Caine at Wikes, who had taken her in, even being baptised; but she was "rather loose and wanton in her conduct (from Ralph of Coggeshall, as translated by Thomas Keightly, 1828)." She asserted that she came from a vast cavern world of eternal twilight, wherein which all the inhabitants were of a green coloration. There was a wide underground river there, which they never crossed, but they could see another (larger?) cavern world on the far side, "more brightly lit." She and her brother had wandered off after their "flocks," and followed the "sound of bells (perhaps distant church bells of the surface world? This "immunity" to "church bells," which are usually anathema to underworld beings according to tradition, could be explained by a mostly-human, albeit abducted, lineage)." They thereby became lost in tunnels which led away from their homeland. Eventually they found their way to the cavern which opened onto the surface world, where "they were struck senseless by the sun, and the unusual temperature of the air." Here we see both unaccustomed solar radiation and air pressure (depressurization?) affecting natives of a deep underground region; doubtless, they staggered about blindly at first, which explains their "disoriented" condition when found. The monastic scholar William of Newbridge, at first skeptical about this account, investigated it extensively and found it to be accurate and truly related in every detail, as the girl had eventually matured into a beautiful woman, and had married a man from the town of Kings Lynn. And their descendants may live among all of us--may even be some us--in the present day.

5. HAIR TODAY, GONE TOMORROW

One of the categories of denizens of the underworld which is most commonly seen (but is of course never actually captured, and scarcely ever caught on film, before mysteriously "vanishing") is the hairy humanoid type, which comes in a variety of sizes, shapes, and dispositions. Often these beings, along with UFOnauts, "Men in Black," "El Chupacabras," and many other strange characters, are announced or accompanied by a foul sulphurous stench, or an ammonia-like odor;, perhaps further evidence of their subterranean origins (an odious characteristic they share with the demons of old).

Current fields of thought surrounding the nature of these creatures vary widely, from proponents of a "natural primate" or hominid theory, to a wide-eyed extreme of a "paranormal, elemental (spirit), or benevolent space-brother" mentality. The problem with both extremes is that they both may have circumstantial or ancedotal evidence to support them, but entire reams of evidence are often downplayed, ignored, or scoffed at with an almost rabid spite. The evidence as to the true nature of these creatures actually exists within most folklore and mythic traditions of the world, and their origin--the place from which they come and to which they return--is definitely subterranean, as can and will be shown.

Those who seek to fit "Bigfoot," "Sasquatch," "Yeti," "Wendigo," "Skinwalkers," and other such creatures into the worldview of a surface-dwelling naturalist are simply ignoring facts which are inconvenient or threatening to their own pet theories. Perhaps they are worried about endorsing evidence which might draw ridicule or dismissal in their direction, obsessed as they sometimes are with being considered "serious scientists" by the academic community at large, and the world in general.

A Sasquatch or Bigfoot, drawn by a ten-year old child.

(See Color Inserts)

The problem with this attitude is that it is not in fact "scientific" at all, as the scientific method calls for an equal examination of all evidence and all theories, which simply does not always occur. Perhaps the "crypto-hominologists" would do well to study the beliefs of native peoples who have long-regarded hairy humanoids as something outside of the "natural" sphere. This does not mean that such beings are paranormal in nature, but rather that they have an origin and nature which is lost in antiquity, and which does not fit comfortably or even rationally into our surface-world ecosystem, or our world-view. Native Americans consider these beings to be supernatural and often subterranean, as do the Sherpas of Tibet and Nepal and other peoples around the world. Medieval European demonology and folklore are filled with hairy "satyr-" type demons, goblins and trolls, all with subterranean origin, and the always-nocturnal "werewolf" also comes to mind. Grendel and his mother, from the Anglo-Saxon epic poem BEOWULF, are another example of this, and in fact incorporated both the hairiness and scaliness of subterranean hybrid beasties, as well as the underwater entrances to cavern-lairs so common to many Naga and Faerie realms. "Braunies" or "Brownies" of Scotland are hirsute and solitary, as are many Scandinavian trolls (who tend to be towering in stature as well), and many other forms.

Other problems with considering all such beings to be surface-world, or "naturally-occurring" creatures are numerous. While physical evidence does exist, primarily in the form of footprints and castings of footprints, witness accounts, recordings of distinctly hominid or anthropoid sounds, and an occasional photograph, film or videotape image, still verifiable "scat" or fecal material (evidence) has not been conclusively presented, although the opportunities to do so have supposedly arisen. Actual specimens, living or dead, have not been produced, yet these creatures pop in and out of not only wilderness regions, but populated and even urban areas, without being cornered, followed, apprehended, or otherwise diagnosed in terms of their origin. A few hairs have been obtained from different sites around the world, usually identified as "primate" but "unknown," and this is indeed evidence for the existence of such beings--but it says nothing as to their place of origin, or primary habitat. What it does prove however, is that they are real, flesh-and-blood animals, and not ghosts or hallucinations, "transdimensional buddies from a happier place," or other such sentimental silliness.

The researchers into this particular type of mysterious being who shy away from "elemental" and paranormal explanations are definitely on the right track, but they need to examine other evidence as well. If the cryptozoologists would start looking a little more closely at the possibility of a subterranean connection, they might discover the true nature and origin of the hairy giants (and midgets, and monsters) quite quickly, after all. There is a definite correlation between increased UFO activity and hairy humanoid sightings, both in North America and in the Himalayas (the Nepalese call some types of flying disks "friendly dragons"), and this has even been discovered in Native American legends, some of which connected the "crazy bears" who live in caves with the "strange moons" which sometimes came to pick them up or drop them off, or the "wild men" with the "star people (who probably are not from the stars at all, as we have seen)". Anecdotal evidence does indeed seem to indicate that such beings--along with a wide variety of other mysterious, "unnatural" creatures--seem to be in our surface world for an unknown purpose or mission. They come into an area, interact with humans or infiltrate a desired location, and then depart after minutes, hours or days, seldom staying long enough for many repeated sightings (although this does happen also upon occasion).

Those who would seek to wrap hairy humanoids in a cloak of "benevolence," "ascendance," or non-aggression must not have studied the anecdotal, witness-supported evidence, or are ignoring whatever they find unpalatable. These creatures do attack people, vehicles, houses and mobile-homes; they have been seen to hurl boulders and logs at human intruders, or to harass them by night as they sleep around a campfire. Tales of these bipedal creatures killing both animals and humans, and eating the victims' livers, also seems to be a recurring theme (the liver is one of the organs also purportedly taken from livestock by the Puerto Rican monstrosity, El Chupacabras, another creature with apparently subterranean origins), and like many other non-human bipedal creatures, they seem to have a taste for road-kill, particularly for dogs. They also are obviously more intelligent than the average ape or other non-human primate, as they hide their comings, goings, and origins exceedingly well, apparently even removing their dead, since not even a recent bone has been found. Proponents of the "natural" (i.e., surface-world ecosystem) theory argue with such quips as "When was the last time you saw a dead bear, or found a wild animal skull in the wild?" and other such gems. Actually, this writer has found both dead animals, and their bones, in the wild, and has a small collection of sun-bleached beaver, bobcat, and gray fox skulls to prove it. Many a large mammal is seen dead on the side of the road, from moose, deer and wild pigs, to coyotes, foxes, and yes, even bears....But when was the last time a Sasquatch was plastered across a highway by a semi-truck, for all to see? Surely an enterprising trucker would recognize the good fortune of such an occurrence, particularly if he had a refrigerated trailer in which to store the remains, and a working citizen's band radio to call for assistance. But this has not happened, because hairy humanoids are by and large not simply another form of ape or hominid which populates the surface of the world. They're just passing through. They're a lot like human beings, but larger. And they're SMART.

In His Book, WILDERNESS HUNTER (1892), President Theodore Roosevelt told a horrifying tale entitled THE BAUMAN STORY which he believed to be true, concerning an encounter between two frontiersmen and a hairy humanoid "goblin-beast." Obviously of the Sasquatch/Bigfoot species, this creature first stalked and harrassed two brave trappers, eventually killing one of them when they became separated. Roosevelt relates:

> *"On reaching the pond Bauman found three beavers in the traps, one of which had been pulled loose and carried into a beaver house. He took several hours in securing and preparing the beaver, and when he started homewards he marked, with some uneasiness, how low the sun was getting. As he hurried toward camp, under the tall trees, the silence and desolation of the forest weighted on him. His feet made no sound on the pine needles and the slanting sun-rays, striking through among the straight trunks, made a gray twilight in which objects at a distance glimmered indistinctly. There was nothing to break the gloomy stillness which, when there is no breeze, always broods over these somber primeval forests. At last he came to the edge of the little glade where the*

camp lay, and shouted as he approached it, but got no answer. The camp fire had gone out, though the thin blue smoke was still curling upwards.

"Near it lay the packs wrapped and arranged. At first Bauman could see nobody; nor did he receive an answer to his call. Stepping forward he again shouted, and as he did so his eye fell on the body of his friend, stretched beside the trunk of a great fallen spruce. Rushing towards it the horrified trapper found that the body was still warm, but that the neck was broken, while there were four great fang marks in the throat.

"The footprints of the unknown beast-creature, printed deep in the soft soil, told the whole story. The unfortunate man, having finished his packing, had sat down on the spruce log with his face to the fire, and his back to the dense woods, to wait for his companion. While thus waiting, his monstrous assailant, which must have been lurking in the woods, waiting for a chance to catch one of the adventurers unprepared, came silently up from behind, walking with long noiseless steps and seemingly still on two legs. Evidently unheard, it reached the man, and broke his neck by wrenching his head back with its fore paws, while it buried its teeth in his throat. It had not eaten the body, but apparently had romped and gamboled around it in uncouth, ferocious glee, occasionally rolling over and over it; and had then fled back into the soundless depths of the woods.

"Bauman, utterly unnerved, and believing that the creature with which he had to deal was something either half human or half devil, some great goblin-beast, abandoned everything but his rifle and struck off at speed down the pass, not halting until he reached the beaver meadows where the hobbled ponies were still grazing. Mounting, he rode onwards through the night, until beyond reach of pursuit."

So much for enlightened nature-spirit theories, as well as the picture of a shy and nonagressive primate! A medieval chronicler might easily have called this creature a demon, satyr, or a werewolf. The word and judgement of Roosevelt should probably be taken with great seriousness, as he was both a great woodsman and hunter, and a great statesman. Roosevelt was apparently not a total stranger to unknown creatures of hellish description, as he also allegedly embarked on several hunting expeditions in search of the infamous "Jersey Devil" of the New Jersey Pine Barrens (yet another potential subterrestrial), which was also hunted by the brother of Napoleon Bonaparte, and which was sighted and fired upon (with a cannon!) by U.S. Naval hero Stephen Decatur.

Hairy humanoids or hominids are often sighted in mountainous regions which have abundant caves and sinkholes, as well as near deep lakes or other bodies of water, as with the Conser Lake Monster of Oregon and the Lake Worth Monster of Texas, and the Thetis Lake Monster of British Columbia. Swamplands are also a favorite haunt, as with the violent and foul-smelling Fouke Monster of Arkansas (made famous in the low-budget psuedo-documentary, Legend of Boggy Creek). Like Grendel, they apparently have an affinity for cavernous places, with both above-ground and underwater entrances, which may account for their habit of following or walking along woodland streams. Grendel was indeed such a beast, and bore more than a slight similarity to the creature in Roosevelt's story, as a passage from Burton Raffel's translation of BEOWULF demonstrates:

Grendel snatched at the first Geat
He came to, ripped him apart, cut
His body to bits with powerful jaws,
Drank the blood from his veins and bolted
Him down, hands and feet; death
And Grendel's great teeth came together,
Snapping life shut. Then he stepped to another
Still body, clutched at Beowulf with his claws,
Grasped at a strong-hearted wakeful sleeper
--And was instantly seized himself, claws
Bent back as Beowulf leaned up on one arm.

A mighty battle ensues, and Beowulf, no person of small stature himself, tears the arm from Grendel's giant body:

Snapped, muscle and bone split
And broke. The battle was over, Beowulf
Had been granted new glory; Grendel escaped,
But wounded as he was could flee to his den,
His miserable hole at the bottom of the marsh,

Only to die, to wait for the end
Of all his days....

Beowulf and his men have next to contend with Grendel's murderous mother, and the hero says:

...let us go as quickly as we can,
And have a look at this lady monster.
I promise you this: she'll find no shelter,
no hole in the ground, no towering tree,
No deep bottom of a lake,where her sins can hide.

Beowulf and his band of heroes journey to the home swamp of Grendel and his mother, where they encounter strange reptilian monsters above the submerged entrance to the trolls' subterranean lair, and kill one of these:

They stood and stared at the incredible wave-roamer,
Covered with strange scales and horrible.

The possibility of subterranean origins for creatures of this type are obvious as well.

Next the mighty Geat (Goth) dives into the swamp, to be captured at first by Grendel's horrid mother, and taken into an air-filled space resembling an arched structure, or "battle-hall," beneath the bottom of the lake, which sounds suspiciously like a limestone cavern. The she-fiend is covered with thick reptilian armor (this type of mixing of "hairy hominid" and reptilian characteristics is sometimes reported in contemporary accounts, but is often downplayed or ignored), and Beowulf's sword has no effect. As with her son, he has to grapple with her, but she pulls a dagger and tries to stab him! So much for her utterly "primitive" nature. Beowulf notices that the room is filled with weapons and other booty stolen from vanquished heroes, so he latches onto an ancient sword massive and sharp enough to penetrate her scaly hide. She dies, and he then goes down dark passageways in search of Grendel, only to find the fiend lying dead in a corner. Both of these monstrous Naga-trolls have finally been destroyed.

**Beowulf and Grendel,
By Donn P. Crane**

(See Color Inserts)

Beowulf confronts Grendel's mother in her lair, as drawn by Donn P. Crane

(See Color Inserts)

The question we are left with is: How different is Grendel's mother from Nimue? One is possessed of unearthly beauty, the other is hideous; but both have their "offspring" in the surface world, doing their bidding, in both cases involving a royal house, i.e., human politics. The birth of Arthur is the result of underworld machinations, and the deaths of Hrothgar's warriors at the hands (and teeth) of Grendel are no different in this regard. Both lake-cavern-dwelling ladies are strong and aggressive female leaders, and both possibly have their origins in a subterrestrial matriarchal society, which goes all the way back to the rebellion and exile of Lilith. The dark Hindu goddess Kali, destroyer and eater of human beings in one of her aspects, is another face to this same dark reality. While their forms, or genetic lineages may differ (yet be inter-related), philosophically they are the same. Nimue had as much mercy on the enemies of Arthur, through her gift of Excalibur and other acts, as Grendel's mother had on the Danes. Another interesting parallel is the "stolen" yet super-powerful sword that Beowulf found in the cavernous lair under the swamp-waters, and used against the she-troll; what would have been the result if Arthur had invaded the realm of Nimue, and seen her as she really was? Would he have used her own weapon, Excalibur, against her? The parallels, and the circumstantial evidence, just don't seem to stop.

The return of Excalibur to Nimue as drawn
by Aubrey Beardsley

The poet leaves no doubt as to the origin and home of Grendel and his mother:

A powerful monster, living down
In the darkness, growled in pain, impatient
As day after day, the music rang
Loud in that hall, the harp's rejoicing....

And

The Almighty drove
Those demons out, and their exile was bitter,
Shut away from men; they split
Into a thousand forms of evil--spirits
And fiends, goblins, monsters, giants,
A brood forever opposing the Lord's
Will, and again and again defeated.

While the Anglo-Saxon author credits the Biblical Cain with part of Grendel's ancestry, the description sounds identical in every way to that attributed to the children of the possibly-hominid Lilith, and her ancient reptilian allies.

The northlands of ancient Europe were by no means the only places haunted by the monstrous male and female humanoids, scaled and hairy, of the wastelands and cavern-realms. In his monumental translation of the great work, THE ARABIAN NIGHTS, Sir Richard Burton elaborates on his description of a cannibalistic Ghulah, (Grendel's mother?) who is able to appear as a beautiful woman (Nimue?) when the need for such deception arises. About the Ghulah, his annotation reads:

"The Ghulah (Fem. of Ghul) is the Heb. Lilith or Lilis: the classical Lamia; the Hindu Yogni and Dakini; the Chaldean Utug and Gigim (desert-demons) as opposed to the Mas (Hill-demon) and Telal (who steal into towns); the Ogress of our tales and the Bala yaga (Granny-witch) of Russian folk-lore. Etymologically "Ghul" is a calamity, a panic fear; and the monster is evidently the embodied horror of the grave and the graveyard." i. p55

(See Color Inserts)

How long will the "Nation of the Third Eye" be pemitted to victimize mankind? And why is this exploitation revered? Is some worship born of fear?

And the "Ghuls," from which our word "ghoul" has come, are a race are no less heartless, predatory, and bloodthirsty:

> *"Allah ease thee, O King of the age even as thou hast eased me of these Ghuls, whose bellies none may fill save Allah!" ix. p. 152*

Proof of the underworld lust for human or hominid genes can be found, if searched for, in nearly every folklore tradition, as can their tendency to see primates, including humans, as a food-supply. The Tibetan Buddhists have a legend which states that the people of Tibet are descended from a "Monkey-King--" i.e., an advanced hominid or human--and a "Rock Ogress," a being which lives in caverns and haunts the snowy slopes of the Himalayas. The ogress attempts to seduce the monkey, at first unsuccessfully, until she plies him with these words:

Alas! Great Monkey King!
Think of me a little and hear my plea.
By the power of my karma, I was born among the race of ogres.
As my lust grows, I have become enamoured of you.
Driven by my desire, I have come to make this request.
If you will not marry me
I will take a rock-ogre husband.
Every day we will slay ten thousand living beings,
And every night we will devour a thousand creatures.
I will bear countless ogre-children
And this snowy realm will be filled with ogre-cities.
Every living creature will become an ogre's prey.
By comparison, is it not better to think of me
And show your compassion?
(From The Clear Mirror: a traditional account of Tibet's Golden Age, Sakyapa Sonam Gyaltsen's 'Clear Mirror on Royal Genealogy', translated from the Tibetan by McComas Taylor and Lama Choedak Yuthok. Published by Snow Lion Publications, Ithaca New York, 1996)

From this union, and to the Monkey's initial woe, Tibetan humans were born. In case the connection is lost on the reader, the "ogres" are synonymous with the Yeti of the Himalayas. While smacking of subterranean propaganda worthy of Merlin or a Japanese Dragon-King, this demonstrates again the universality of the same concepts previously seen.

Returning to the connection between hairy beings, water, and caverns, the well-known UFO abductee Betty Hill, in an interview with author and bigfoot-researcher Jack Lapseritis (who is a proponent of the "benevolent space-brother" Sasquatch theory), gives an excellent example of the strange connections which exist in our modern, waking world. To quote from his book, "THE PSYCHIC SASQUATCH," Mrs. Hill told him the following about her family's knowledge of Bigfoot in New England:

"Another strange thing happened on the Connecticut River. I don't think there was ever any publicity about this, although people who lived in the area all knew about it. My informant said that at one spot in the middle of the Connecticut River there was a good-sized island that was uninhabited. Then

suddenly one day it was inhabited by small, prehistoric-appearing (hairy?--W.M.M.) people. They didn't know how many of them there were, maybe 50. They lived on the island for three years. No one ever succeeded in getting near them. The police had gone out to the island on boats and had gone onto the island. These small, primitive people could outrun anyone. They would take off running and then could not be found. There were many caves on this island, and it was believed that maybe they were hiding in the caves, but the police just could not find them. Anthropologists went to the island and could observe them at a great distance but couldn't get close to them. No one ever succeeded. It is not known how they lived or what they did for food. No fires were ever seen on the island, but they lived there year-round for approximately three years. Then, just as suddenly as they had appeared, they disappeared."

(See Color Inserts)

A samurai fighting a KAPPA and aquatic reptilian humanoid.
Illustration by: Utagawa Kuniyoshi

She then relates how a helicopter tried to get a close look at these four-foot tall, "prehistoric-looking" people, but they would out-run the helicopter at an amazing rate of speed and "just disappear." The sightings of these "mystery hominids" coincided with ongoing UFO activity in the same area.

Lapseritis' book contains many additional accounts of great interest (although the New Age slant regarding the temperament and nature of Sasquatch is rather heavy), including Sasquatches guarding mysterious cave entrances from intruders, underground machinery noises associated with bigfoot sightings, and sightings of hairy humanoids and "gray" aliens together. If the Sasquatch is indeed a watchdog, errand-runner and hominid servant of an even more mysterious and less-seen race or species, then this would indeed explain much about the behavior of these hairy giants (or midgets, as the case sometimes turns out to be).

Their ability to endure the extremes of nearly all surface-world environments, including sunlight, might make them valuable servants of hidden masters. Primates, or primate-hybrids, as slaves to ancient reptilian intellects? Sounds far-fetched, but many clues exist to indicate that this is the case, from hairy trolls, demons, and other medieval tales, to all of the accounts mentioned in this section. This does not mean that some "hairy humanoids" could not be unknown anthropoid apes or primates as yet undiscovered by science, and living in the wild; perhaps a few of these do continue to exist in a low-profile way, as the gorilla did until the late 19th and early 20th century. But the sheer variety of types seen and described and the frequency of these sightings (even in populated areas), along with the peculiarities of behavior (furtiveness, "focused" on an apparent mission at times) and other circumstantial evidence, indicates that the majority of such unknowns are quite possibly affiliated with a vast, unseen, yet natural, reality.

(See Color Inserts)

6. THE FARMERS IN THE DELVE

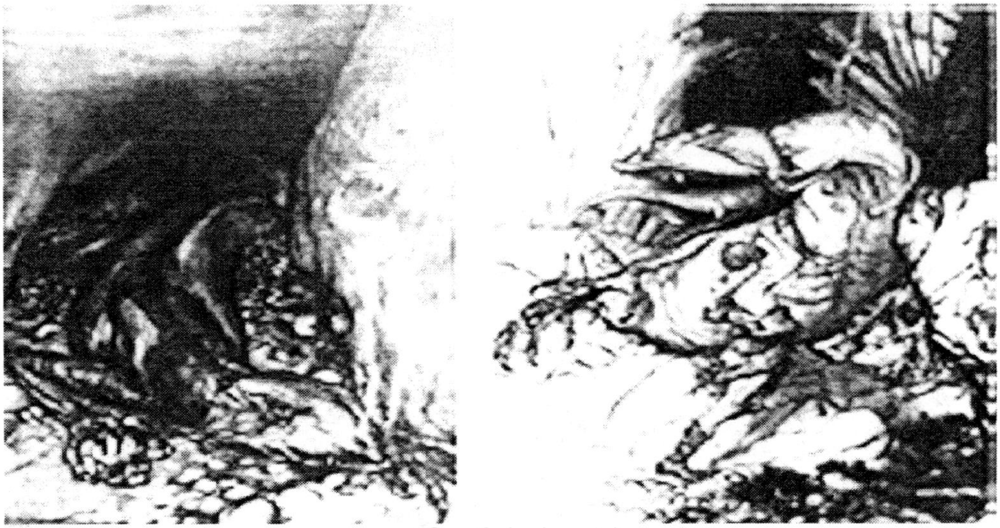

(See Color Inserts)

**The nightmare of every underworld being.
To be discovered is often to die.
Not that they don't have it coming.
Illustration by: Arthur Rackham**

Rumplestiltskin and Tom-Tit-Tot longed to steal a human child; the Pied Piper of Hamelin took more than a few. Demon lovers, djinn, succubi and incubi haunt our collective unconscious. "Fairy princesses" and "frog princes," somewhat human, somewhat other, populate the world of pre-adolescent folktales, as do any number of stunted, deformed, or miniscule beings. Dragonsand monsters lurk in the dark corners of our minds, and always have. Some claim to still see, against their will, such beings in dark lochs, thick woods, and deep cellars and caverns, or coming and going in flying, glowing chariots or disks. Are the paranormal, cryptid, and fortean forms of today simply the stuff of which folklore is made--or is it the other way around? Are these things, these beings and their activities, the "ground of being" from which all of our mysteries and terrors have actually sprung, and continue to spring?

The victim of a traumatic event sometimes finds that his or her memory of that event is not perfectly clear; often it's hazy, disjointed, and comes back in snatches of memory and images which jolt one from sleep. Other victims of trauma, abuse, and terrifying encounters or experiences find that the truth of what happened to them has been covered up by confabulation, and upon occasion by a blanket of lies woven by the perpetrator or perpetrators.

Could it be that our entire folklore and mythic experience is just such an confabulation, but on a species-wide scale? Has humanity been conditioned, are we being conditioned, to overlook and dismiss, to turn into 'fairytale" and "fable" very real and ongoing events and victimizations? Are we simply a "resource" to somebody, used and deceived by an ancient enemy which occasionally masquerades as friend, when it serves a purpose? Do humanity and the rest of the surface world represent a genetic smorgasbord or farm, where genetics, protein, and other goods and services are readily available, and are at the command and demand of an unseen population of diverse races and specie?

All of the religious, folklore, and mythic traditions must surely not be, as Joseph Campbell once postulated, the mindless result of the "energies of the body," the "activity of the organs." Humanity is not the cause of reality, but is a part of reality, and a largely-ignorant part at that. The farther one is removed in time from originating events, the dimmer the truth becomes--to the animals in the field, that is. The farmer and the hunter know full-well where the animals came from, what their lineage is, and how to take them down for any purpose at any time. The animal reacts as an animal always will, with fright, confusion, aggression or flight--but the hunters and farmers are always still there, somewhere behind the scenes, in their hunting-blinds or farmhouse parlors, accustomed to their role and ready to act with impunity. The final questions arise:

What are the folktales of the beasts of the field? Is humanity the helpless livestock and fodder of an older, predatory race or races? And if we are, will we always permit it to be so?

(See Color Inserts)
Children Going Underground

(See Color Inserts)

Part III:
ECHOES of the WORLD CAULDRON:
LEGENDARY, MYTHIC AND RELIGIOUS BELIEF IN LANDS BENEATH THE EARTH

(See Color Inserts)

Humanity has always been fascinated by the mysterious and unexplored. The modern speculation on the possibility of flinging ourselves into the vacuum of interplanetary space is identical to the drive which drove the Phoenician, Viking, and Polynesian peoples across tremendous expanses of hostile ocean in search of new frontiers, and that which sent the horsemen of the steppes-- Scythians, Huns, Mongols, and others--on their periodic migrations of flaming, sharp-edged pillage. It seems that, in the human psyche, and particularly as reflected in mythological and folklore treatment, those places and rewards most distant and won with the greatest difficulty, have always seemed to be most worthy of winning and settling--or at the very least, of exploring in a quest for knowledge. Folklore, mythology, and the religious texts of the world are replete with tales of mysterious and even hidden lands, such as the kingdom of Prester John, and the Seven Cities of Gold for which the conquistadores so vainly sought. The mythical places were often those where the laws which governed the known world were not always in effect, where Time itself did not always pass in a respectable, predictable fashion as it did in the everyday world. Those mythic realms were never easy to reach, always hazardous to dare, and often impossible to leave; and only the staunchest heroes could safely make such a journey, and return to tell about their explorations.

Foremost among these realms were three distinct regions, called by many names but universal to nearly all cultures (or at least to those who had reached a certain degree of sophistication or achievement, in terms of boldness of migration and exploration). This triple hegemony consisted of fairyland/faerie, where the magical inhabitants dwelt in a timeless state, untouched by the seasons or by the passing of day and night; cavern and tunnel realms of varied scope, peopled by giants, dwarfs, monsters, and a variety of diminutive humanoid types; and the Underworld proper, the deep and immense land of the dead and of the imprisoned, which stretched horizonless and gloom-filled, somewhere far beneath the daylit world. In many regards, these tales were strangely similar or even identical. Often these similarities were in regard to the location of the magical or mythical subterranean kingdoms, the identities and dispositions of the various inhabitants, and the explicit directions as to how to "get there," and then back home once again. Perhaps due to problems of their remoteness, the real or imagined caverns of the world, the deep reaches of the sea, and the ultimate northern areas of the planet were most often chosen as the locations for the mysterious regions where one could physically pass from this world and into another.

(See Color Inserts)

In many accounts these realms touched upon or bordered one another, or were separate provinces of a common kingdom, with Faerie bordering on the land of the dead, or the subterranean grottoes of other ancient races bordering or interconnecting them all. This was especially evident in the mythological worldview of the Norsemen, Celts, and Classical Greeks and Romans; while the northern mythologies were vastly different from their southern counterparts in a fundamental, philosophical sense (compare, for instance, Odin's bloody fatalism and mortality, against the lackadaisical escapades of a philandering and invulnerable Zeus), in many details their underworlds were strangely similar. Almost certainly some cross-contamination or exchange occurred, particularly between the Romans and the Celts of Gaul and Britain; but often the similarities of these legends to those of other cultures which could never have known of one another were uncanny. In the Greek view, the details were quite complete. Dividing the lands of the living from the region of Tartarus, the land of the dead, were many fierce obstacles, rivers, and bodies of water or fire. The greatest of these was Oceanus, or Ocean, which not only comprised all the seas of the world, but was also the largest of the "rivers" which the Hellenes believed swept into and through Tartarus, to emerge from the underworld on the opposite side of the Earth. Another of these subterranean torrents was the Lethe, river of forgetfulness, while another was the Styx, the river of death, which bordered Tartarus and was all but impassable. According to The Odyssey, the Styx was the subject of a mighty oath sworn to by the untrustworthy Gods themselves, an oath which even they would never break:

> *"Now, by the down-flowing waters of the Styx--the greatest and most dread oath of the blessed gods--"*

A map drawn by explorer and cartographer Martin Frobisher. showing an inexplicable "unknown land" at the top of the world. the shoreline of which he had attempted to detail from his ocean voyages.

Deep in Tartarus was the sludge-filled Acheron, dark and cthonic, as well as the river Cocytus, and the fiery, magma-filled river Pyriphlegethon. Many of these rivers had exact underworld counterparts in the beliefs of other peoples.

(See Color Inserts)

Caverns, Cauldrons and Concealed Creatures

According to the Greeks, all the waters of the Earth would sooner or later find their way into the yawning gulf of Tartarus, to eventually emerge again on the opposite side of the world. Tartarus was said to "sink twice as far below the Earth as the Earth was beneath the Sky;" it was bounded by many perils, such as a wall of midnight darkness, a fence of bronze beset with traps and serpents, and the various perils of its attendant waters: clashing rocks, whirlpools, and flows of fire and ice. And of course there were the guardian beings: the serpent-haired gorgons, the hideous harpies, the serpent-maned, multi-headed hound Cerebus, and others. Deep in the depths of Tartarus were the imprisoned, most-feared of its inhabitants, the defeated, ancient, and dethroned gods called Titans, and their monstrous children; they had been thrown into the dungeon-like gulfs of the underworld, hopefully never again to emerge into the sunlit world. Tartarus also contained a variety of regions or kingdoms, ranging from the twilight "fairyland" of dim pleasure, the Elysian Fields, to the many grottoes, caverns and pits of torment which were reserved for the damned and despised.

The Norse cosmology was strikingly similar, but represented differently in its cultural context. The Scandinavian and Germanic peoples envisioned the world--their universe--as an immense yew or ash tree, the limbs and roots of which spread into a variety of worlds or planes of existence. This all-encompassing structure, the World Tree Yggdrasil, plunged its roots deep and into several subterranean kingdoms, which all bordered a vast gulf called Ginnungagap, meaning "Seemingly Empty." This vast and primordial void preceded the creation of all of the realms on the tree, and was an ancient hollowness, immense beyond description. One root of Yggdrasil led into Niflheim, the land of the dead, which contained the kingdom of Helheim, a region ruled over by the horrid death-goddess Hel. Close by this root sprawled the giant serpent-dragon Nidhoggr, who fed on the unfortunate dead in the afterlife. Nidhoggr also gnawed at the root of Yggdrasil, the axis of the world itself. Here the monstrous hound Garm, equivalent to the Greek Cerebus, sat huge and bloodthirsty, as guard-dog. As in the Greek Underworld, many waters flowed out from the depths and into the world of men; in Niflheim, it was the spring/river Hvergelmir, which boiled and churned relentlessly. Hvergelmir meant "Roaring Cauldron," and cauldrons are one of many symbols of the underworld which are found again and again in this genre of myth and folktale. This image of the underworld as a cauldron was often synonymous with, or related to, the womb/maw of the Earth Mother/Death Goddess herself. From Hvergelmir flowed the Gjoll, which surrounded Helheim like a moat, or a cauldron's lip. This is perhaps more meaningful than is at first apparent. The raging Hvergelmir was the source of eleven rivers in total: in addition to the Gjoll, there were the Svol, Gunnthrun, Fimbulthul, Fjorm, Slid, Sylg, Hrid, Vid, Ylg, and the Leipt. It is interesting to note the similarity between the name of the river Leipt, and that of the river of forgetfulness, the Lethe, from the Greek picture of Tartarus. The eleven tributaries of the Hvergelmir emptied into the central void of Ginnungagap.

The second of Yggdrasil's roots found it's way into the lands of the Gods, Asgard and Vanaheim. Again we find a wellspring or flowing water source, this time in the form of the Well of Urd, known as the Well of Destiny. Here dwelt the gods, and the three Norns of the Well, the latter synonymous with the Greek Fates. While often pictured in a poetic sense as a land lofty and high in Yggdrasil's branches, this realm was in fact a subterranean one as well (as it's location by root and well attest). In fact, the only world of the Norse cosmology which can be said to be non-subterranean is that of Midguard (Middle-Earth), the surface world of human beings. Bifrost, the "Rainbow Bridge" which stretched across Ginnungagap and into Asgard from Midguard, may have been attributable to the imagined rainbows created where the waters of the surface world fell away to crash into the pit of the underworld, a place of mists and clouds.

The third major root of Yggdrasil delved deep into the eternally-frozen land of Jotunheim, the land of the Frost Giants, enemies of both gods and men. Here were the waters of the well/cauldron of the giant Mimir, which was called the Well of Wisdom. Here Odin, king of the Norse Gods, obtained one long draught in exchange for one of his eyes, and nine days and nights speared to Yggdrasil itself. The frost giants, symbols of the monstrous, collective Id, ruled Jotunheim with strange magic, illusion, and a deep, dark wisdom which had been lost even to the gods.

(See Color Inserts)

Other subterranean countries were Alfheim, a land of eternal twilight, and home to the unpredictable but often mild-tempered elf-folk; and Svartalfheim, the kingdom of the dark elves or dwarves , a stunted, hairy, unfriendly race which was nevertheless quite clever in terms of technological skills. Nidvellir was another kingdom of the dwarves, perhaps not as negatively-disposed toward humanity as those of Svartalfheim. Muspelheim, land of the fire-giants, was also reached via Yggdrasil, but was separated from all other reaches by the width of the great vastness of Ginnungagap. Muspelheim was located in the far southern reaches of the underworld realms.

(See Color Inserts)

In the Elder Edda, Odin says:

> *"No one has ever known or will ever know the vastness of the roots of that ancient tree."*

This is not only a reference to the created world and heavens as seen by the Viking peoples, but was equally a reference to the root-like cavern system which they believed underlay the surface world--in fact, it sounds more like a reference to the subterranean realm than to anything else. Also issuing from the depths of the World Tree was Jormungand. the titanic world-serpent/ouroboros which "encircled the Earth" and held his tail in his teeth. His writhings beneath the sea were one of the sources of storms and earthquakes. He was called by the Norse skalds "the Girdle of the World." And imprisoned in the depths of Niflheim was the fallen trickster-god Loki, father of Hel, Jormungand, and other monstrosities, who awaited his release and the beginning of Ragnarok, the Battle of the End of the World.

Here we see the chained titan and the dethroned devil; while in Loki's child Jormungand, we see a theme just as common to the underearth cosmology--the serpent/dragondevourer, present in Gorgons and Cerebus (both serpent-haired) as well as in Nidhoggr and Jormungand. As will be seen, these motifs are common elsewhere as well. It is the account of Ragnarok, at which the Aesir and Vanir Gods perish in battle against the unleashed denizens of the underworld region, that there finally emerges a definite and concrete clue as to the primary location of the entrance to those buried realms. Out of the farthest North comes the ship Naglfar, built of the fingernails of dead men. and carrying Loki, Hel, frost giants, and others. This ship is driven by the thrashing of the circular Midgard Serpent (who therefore must also lie in the far north), in the depths of the northern seas. Out of the ultimate south comes the fire giant Surt, lord of Muspelheim, the subterranean realm of fire. Gods and men are caught between these polar forces of unleashed evil, and nearly all of the combatants are destroyed. Like the underworld of the Greeks, the Norse version had two entrances as well. In keeping with the theme of a northern location for the entrance to the underworld, the Greeks believed that one of the entrances to Tartarus lay beyond "Hyperborea," meaning "Beyond" or "Behind the North Wind." They and the Romans referred to the Arctic regions as "Thule," or "Ultima Thule," the "ultimate" northern lands of the Earth. One wonders if the Norse underworld river "Fimbulthul" did not run through this mythic region....

Other peoples, such as the Finno-Ugric tribes, held similar beliefs. Living at the edge of the Arctic as did the Norse, the Finns regarded the hostile, foreboding northernmost lands with superstition and fear. They saw it as a region of dark sorcery, particularly the area of Lapland, which stretched northward into the Arctic Circle, and to the shores of the polar sea. Beyond the land of the Lapps was the entrance to the Finnish underworld and land of the dead, Tuonela, ruled by the lord of the dead, Tuoni. In the Kalevala, the national epic of Finland, there are several references to this morbid clime, similar in many respects to the underworlds of other cultures, complete with turbulent bodies of water, whirlpools, rivers of fire, and monstrous serpentine entities, both aquatic and terrestrial:

Into Tuoni's steep rapid,
into the frightful eddy
down the highest waterfall
down into the worst whirlpool.

And:

That was Lemminkainen's end
The untiring suitor's death,
Down into Tuoni's black river
In the Dale of the Dead Land.

Again:

Your son, luckless you
Has been lost, been killed
Down in Tuoni's black river
The Dead Land's ageless water--
Gone through the rapids roaring
With the currents in a flash
Towards furthest Tuonela
To the Dales of the Dead Land.

Or:

He led me to Death, pushed me
into Tuonela's river.
He raised a snake out of the water
A dragon of the waves
Against woeful me--
From the sky down to the earth;

Here are references to the torrents of Tartarus and Ginnungagap, and the fierce dragon-guardians of the dark lands.

They sought me in Tuonela
the Dead Land, underground too--

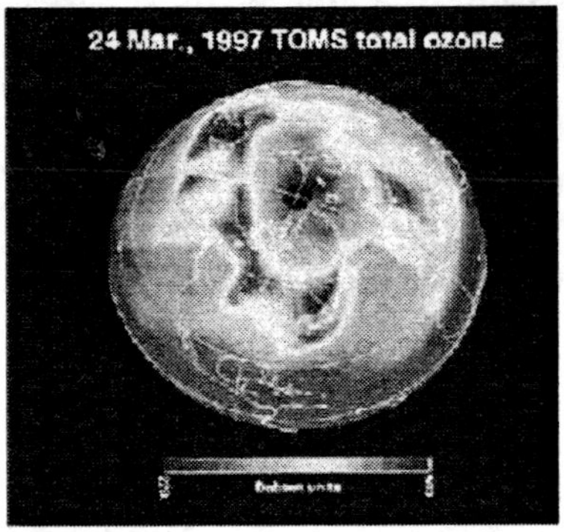

The two top NASA images at left show the "ozone depletion" which seems to always exist at the poles, where the atmosphere is said to be much thinner (although as a reegion of frigid temperatures, it should instead be expanded upward – unless there is something about the surface of the earth there). The second ultra violet image shows both North and South polar regions, where sloping 'indentations' in the atmosphere are evident.

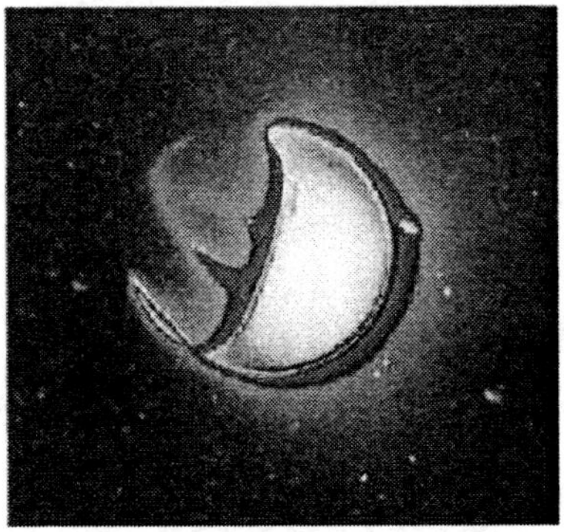

(See Color Inserts)

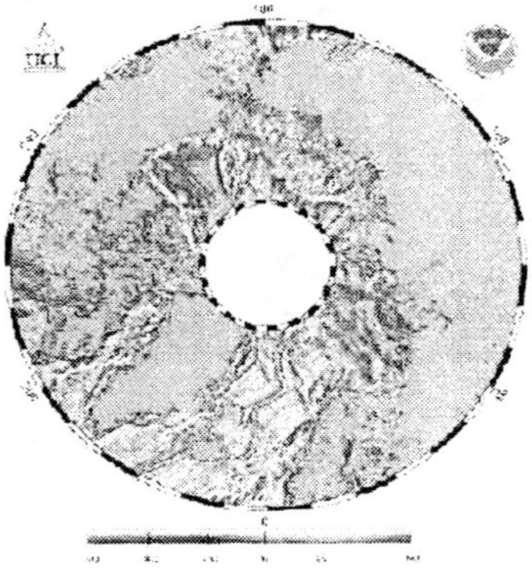

At bottom left is gravitational data of the North Polar region, from the UCL-NOAA Arctic and Antarctic Research Center. Like most satellite images of the North Polar region, most of their data which is actually in the vicinity of the pole is missing, due to the fact that most satellites do not actually pass over the geophysical and magnetic poles, but skirt the edges. The satellite which created this miage only reaches to 81.5 degrees north, and measurements are only obtained from directly beneath the satellite.

And a christianized pagan reference:

May the Good God not
May he not bring this about--
He who himself went to Death
Penetrated Tuonela.

As in other mythologies, the entrance to Tuonela was that mysterious place where sky and earth met in some awesome configuration which existed nowhere else:

You'll travel a tiny way
And you'll come to Northland's yard:

An iron fence has been built
A steel pen has been fashioned

From the earth up to the sky
From the sky down to the earth;

This barrier is guarded by serpents and 'worms," the latter not being invertebrate annelids, but huge and terrible dragons ("wyrm" and "worm" often meaning a dragon which dwelt within or beneath the earth). Tuonela is filled with these monstrosities, as well as with witches, sorcerers, and demons. Lava-filled whirlpools and rivers also abound. Before reaching the place where sky and earth meet, one must cross a frozen sea:

"To the Edge of the North Sea,
To the Brink of the Boundless--"

Here is the great gulf of the underworld, where Oceanus rushes into the immeasurable concavity of Ginnungagap. There are also references to the monsters which lay chained down in the darkness, as the good wizard Vainamoinen bound one of them himself:

Then the old Vainamoinen
Threw the rascal into the waves
And he put this into words:
"Look monster, the Gaffer's son,
Don't you rise out of the sea
Don't come up from the billow,

In front of mankind
From this day forward!"

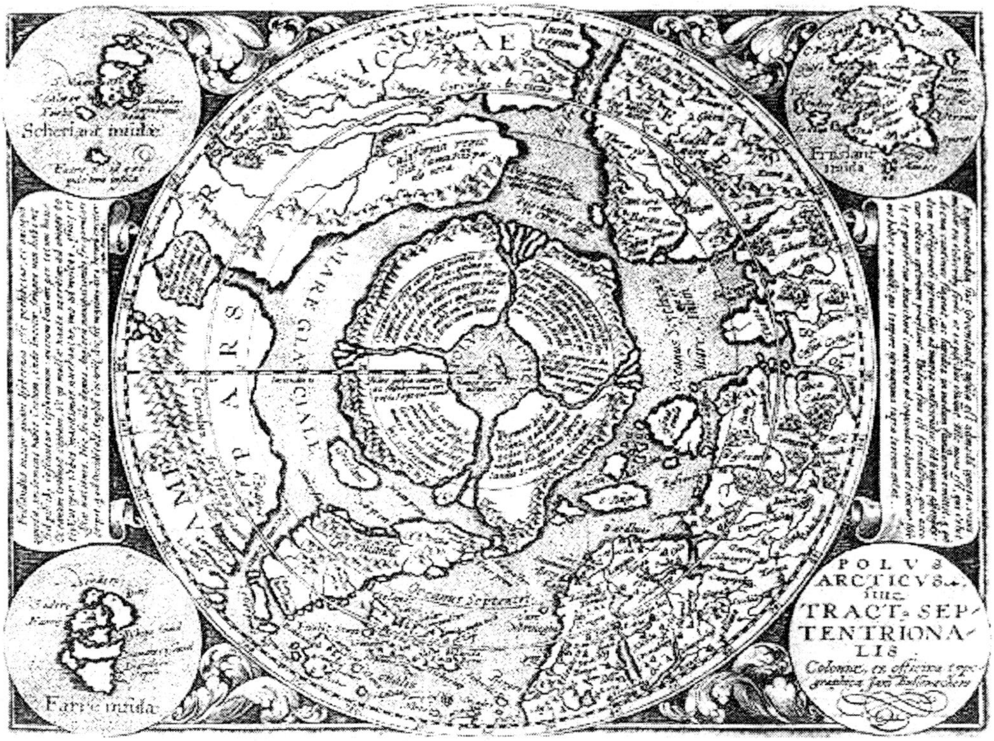

Another polar map showin an unknown land, complete with rivers, at the roof of the world
(See Color Inserts)

So again was Jormungand banished beneath the waves of the cauldron's rim, and a Titan was contained until doomsday at the edge of the polar abyss.

The Eskimo, Inuit, and related peoples of the Arctic region have their share of similar legends. The Eskimos' creation myth states that they came out of a land in the ultimate north, from a subterranean realm where there was only one season, and eternally-good hunting. Their ancestors were "flown" out of this netherworld by their creator/trickster-god, Raven. In

Northern Tales (Pantheon Books, 1990), page 83, folklorist Howard Norman recounts an oral tradition taken from the Greenland Eskimos, entitled "What is the Earth?" In this story, three kayaking friends were "curious about the size and shape of the earth." In their journey they came upon "an enormous icehouse (igloo)," and entered it. The interior seemed to be "without end." They wandered inside for years, until two died of old age and exhaustion. The lone survivor found his way out again, and upon returning to his people as a very old man, he told them:

> *"The Earth is simply a very big ice-house (igloo)."*

So the Eskimos, as well as the Finns, stumbled into Ginnungagap, from which few return.

In Northern Tales, there are also oral traditions recalling Wooly Mammoth, passed down for thousands of years (or were they?). One of these (pg. 271) is called "Why Wooly Mammoths Decided to Flee Underground." In this tale, pressured by over-hunting, the elephants evade humanity by retreating to a subterranean realm beyond human reach. This brings to mind not only the remarkably well-preserved (freeze-dried) remains of deceased mammoths which surface from time to time from the ice and permafrost in Alaska and Siberia, but also the cryptic account allegedly given by an astonished Admiral Byrd, after one of his south polar flights, when he claimed to have seen (and filmed) a mysterious animal which looked like a wooly mammoth, and to have flown to, as he called it:

> *"That enchanted continent in the sky, land of everlasting mystery."*

As for the north polar region, Byrd claimed to have flown to an extent of 1700 miles beyond the pole, and into an unknown land-- a land where surface topography could be seen to "hang in the sky." Before leaving on his seven-hour flight over the north polar region, Byrd said:

> *"I would like to see that land beyond the pole. That area beyond the pole is the center of the Great Unknown."*

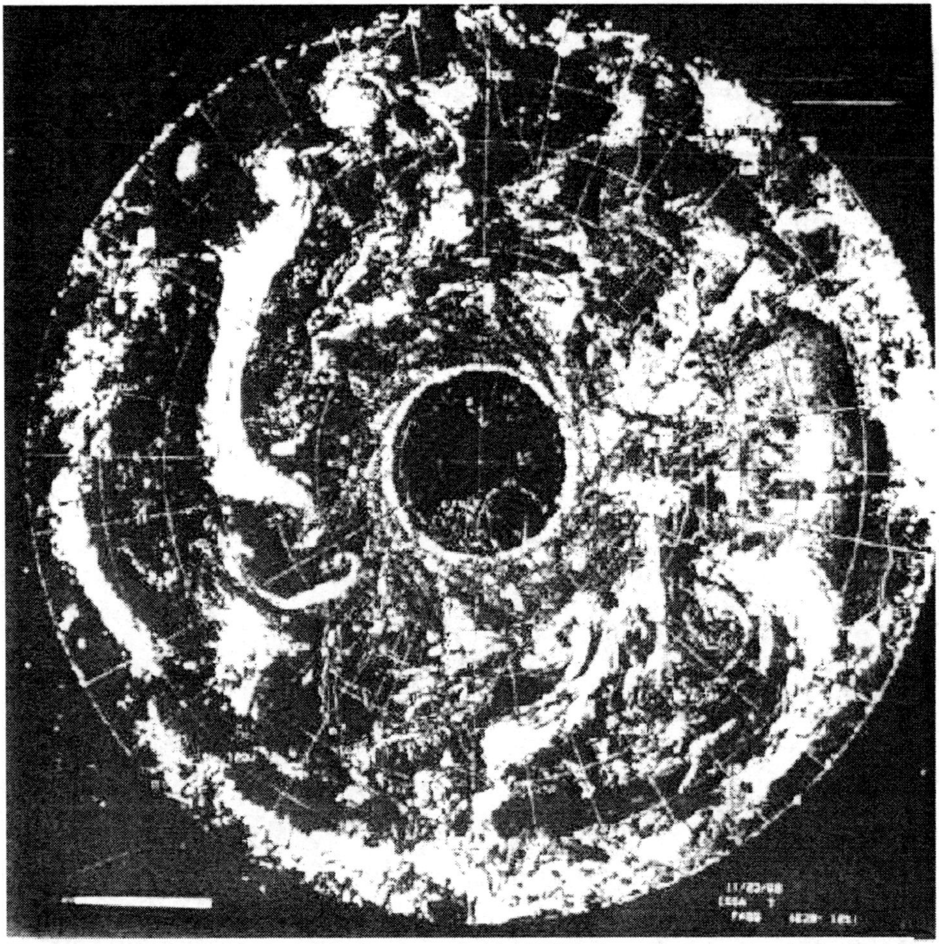

The controversial photo image from data recorded by ESSA-7 satellite, showing the north polar region on November 3rd, 1968. Conventional explanations, such as 'polar night' or shadow, only fail serious consideration, particularly since not only is the line of demarcation exactly idenitfiable, but attendant 'cloudy' or weather forms can be seen going over the 'lip,' and within the darker 'void' itself. The enigmatic 'shifting' of the disputed geophysical poles may be the result of reading taken along the rim itself, where a sharper curvature of the surface – as noted by Byrd and others – has begun. Perhaps someone has indeed 'described a circle upon the face of the waters.'

Photo: Environmental Science Administration,
U.S. Department of Commerce;
Flying Saucers magazine (Ray Palmer)

(See Color Inserts)

Byrd almost certainly was not referring to Scandinavia, Russia, or Siberia, as their locations--and names--were quite well known at the time. To what region, then, did he refer? The "center of the Great Unknown" sounds eerily similar to the Norse Ginnungagap, with its uncharted vastnesses stretching beyond the roots of the World-Tree.

Again, in "Northern Tales," we find a Chukchee tale called "The Attainable Border of the Birds." This starts with an account of a "monster-filled" sea to the far north, behind which is the land "where the solid sky falls down, striking against the Earth." This barrier opens and closes periodically, sometimes killing flocks of birds in the process. Here is an account of a strange phenomenon almost identical to that related in the Kalevala.. Within the confines of this strange boundary is the land to which the Arctic birds migrate in the winter time. This "migratory disappearance" is a very real mystery which still confounds both the indigenous people of the region, and scientists.

Many Native American tribes have myths which claim that their people originated in a subterranean realm, often driven out by reptilian or other unhuman persecutors. This legend is widespread, from the Northwest to the Northeast of the United States, and from the tribes of the Southwest--Hopi, Apache, and others--to those of the Southeast, in particular the Choctaw. In many Indian traditions, the Earth is seen as a great, hollow gourd or nut, with many things living both within and upon it.

According to Aztec origin myth, they originated from Chocomortoc, (the seven caves), from which came the legend of the 'Seven cities of Gold.'

(See Color Inserts)

As mentioned previously, another parallel which exists among many traditions is that of the Earth and it's underworld pictured as an enormous cauldron, filled with the mysteries of life, death, and eternity. The underworld is also the source of magical cauldrons, perhaps the original cauldrons which found their way into the folktales about witches and their brew. These magical cauldrons were "horns of plenty," cornucopia of magical elixirs and beverages, or of undead legions ready for battle. In the Norse stories, we find the thunder-god Thor journeying in disguise into the land of Jotunheim, to secure a mystical beer-brewing cauldron from the giant Hymir. Not only did Thor obtain the cauldron, but as a bonus, he and Hymir made a fishing trip in the waters verging on Jotunheim. Here Thor hooked into the Midgard Serpent himself, and a furious battle was joined--but before Thor could hoist it into the boat, a terrified Hymir cut the line. Thor and the Midgard Serpent would not meet in conflict again until Ragnarok, where both were destined to die.

In the Kalevala, a hero must obtain a mead-making cauldron, or a mead-horn variant:

> *Bring a trump from beyond, from*
> *The pole of heaven yonder*
> *Bring a honey-trump from heaven,*
> *A mead-trump from mother earth;*

The "pole of heaven" is an interesting touch, as this is almost certainly a reference to the "shaman's pole," a variant of Yggdrasil itself, and a common theme in all circumpolar cultures. The central tent pole was generally the device by which the astral form of the shaman would climb to other planes of existence, and the greatest such pole was believe to be at the top of the world, in the vicinity of the Great Bear, and topped by the Pole Star. Put the tent-pole/World Tree in modern context, and you have the star-topped Christmas Tree, a pagan trapping whose mystical roots are buried deep in the collective unconsciousness. It would be interesting to find out if the pagan holiday observance of "Yule" (later Christianized) has its etymological origin in the same root which gives us "yew--" Yggdrasil, after all, is often said to be a yew tree. In the Celtic tradition, cauldrons are of course associated with the underworld. In the Mabinogion, the land of Annwn (literally meaning "unplumbed," or "bottomless") which is the Welsh underworld, contains a mystical cauldron which can restore the dead to life once more, if they are submerged in it and brought out again. This is used to create an "Army of the Undead," for these beings are no longer truly human--they are automatons, to be killed and revived again and again.

(See Color Inserts)

This is symbolic of the murky underworld realms, the strange concavity believed to lie in the ultimate north by many cultures, which bubbled like a stew-pot with torrents, whirlpools, steam, and ice, and into the depths of which all of the dead were believed to eventually go, like so much stew-meat (This also calls to mind certain modern variants of UFOlogy folklore and internet rumor, regarding the genetic-engineering experiments of subterranean-based "aliens"-more likely physical/paranormal entities-who, according to supposed government "informants," are building hybrid beings from the tissues of other creatures, utilizing "vats" [cauldrons] to melt bodies of livestock and humans down into usable materials). The underworld was seen as a boiling crucible of creation and destruction; little wonder that the classical "rivers of the dead" plunged into its depths, to emerge again as the life-giving waters of the sunlit world.

But the pagan world of antiquity was not the only place where such ideas abounded. The Bible is also filled with similar motifs. In the Judeo-Christian tradition, the underworld, originally called Sheol in Hebrew ("Hell" is actually a variant of the Norse/ Germanic "Hel," the name of the death-goddess of Niflheim; the translation is courtesy of medieval Europeans), is the abode of restless and unclean spirits, the spirits of the dead (in the Old Testament/Torah in particular), and the fallen angels or demons. A dark place of punishment, it contains a pit called the "deep" or the "abyss," (Luke 8:31) which demons beg to not be cast into by Christ. The concept known as the subterranean Lake of Fire is destined to be created and used as a place of eternal punishment and confinement only after the Day of Judgment, and sentencing, has come. The only protection against these roaming spirits of evil was the covenant that one had with God:

> *Have respect unto the covenant, for the dark places of the earth are full of the habitations of cruelty. --Psalm 74:20*

The Biblical underworld is presented as a place of imprisonment and misery, and a pre-destined place of detention for Satan and his demons. They come and go from the deep places of the Earth at present, but one day will eventually be bound forever, in a lightless, inescapable lake or pit of fire or magma. The devil's place in the underworld is referenced in this regard in Isaiah 14:12-15:

> *How art thou fallen from heaven,*
> *O Lucifer, Son of the Morning!*
> *How art thou cut down to the ground,*
> *Which didst weaken the nations!*
> *For you have said in your heart,*
> *I will ascend into Heaven;*
> *I will exalt my throne above the stars of God;*
> *I will sit upon the mount of the congregation,*
> *IN THE SIDES OF THE NORTH;*
> *I will ascend above the heights of the clouds;*
> *I will be like the Most High."*
> *Yet thou shalt be brought down*
> *To Hell, to the SIDES OF THE PIT.*

In Revelation 9:1-3, we find:

> *And the fifth angel sounded, and I saw a star fall from heaven unto the earth:*
>
> *and to him was given the key of the bottomless pit.*

And he opened the bottomless pit; and there arose a smoke out of the pit, as the smoke of a great furnace; and the sun and the air were darkened by reason of the smoke of the pit.

(See Color Inserts)

And from Revelation 9:11:

> *And they had a king over them, which is the angel of the bottomless pit, whose name in the Hebrew is Abaddon, but in the Greek tongue hath his name Apollyon.*

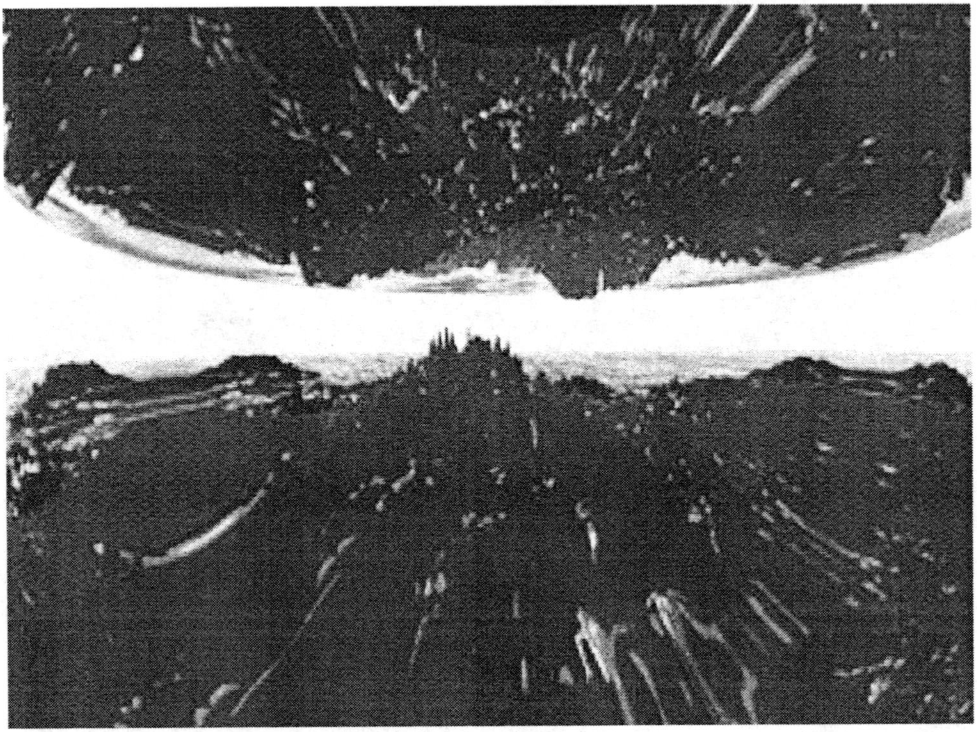

(See Color Inserts)

Here is a parallel with the Greek Hyperborea (meaning "Behind the North Wind"), a mysterious land which Apollo was said to prefer as a dwelling-place, and within which the true "omphalos" or navel of the Earth was located. Apollo was seen by the Greeks as the god and lord of the realm of Hyperborea, a place of eternal luminescence.

Revelation 20, verses 1-3, states:

And I saw an angel come down from heaven,
having the key of the bottomless pit and a great chain in his hand,

And he laid hold on the dragon, that old serpent,
which is the Devil, and Satan, and bound him a thousand years,

And cast him into the bottomless pit, and shut him up, and set a seal upon him,
that he shall deceive the nations no more, till the thousand years should be fulfilled;
and after that he must be loosed a little season.

This release precedes the Judgement, and the final incarceration in the Lake of Fire. Here we find the dark pagan god Loki, the dragon-serpents of Niflheim, and rebellious Titans, all rolled into one--or these are perhaps reflections of a darker, singular truth as related in Revelation/Apocalypse. This is also expressed in Jude 6:

And the angels which kept not their first state, but left their own habitation,
he hath reserved in everlasting chains under darkness unto the judgment of the great day.

And 2 Peter, verse 4:

For if God spared not the angels that sinned, but cast them down to hell, and delivered them into chains of darkness, to be reserved unto judgment-

In the Apocryphal Book of Enoch, Chapter XL, Verse 9, we find:

And I measured out the whole earth, its mountains, and all hills, fields, trees, stones, rivers, all existing things I wrote down, the height from earth to the seventh heaven, and downwards to the very lowest hell, and the judgment-place, and the VERY GREAT, OPEN, AND WEEPING HELL.

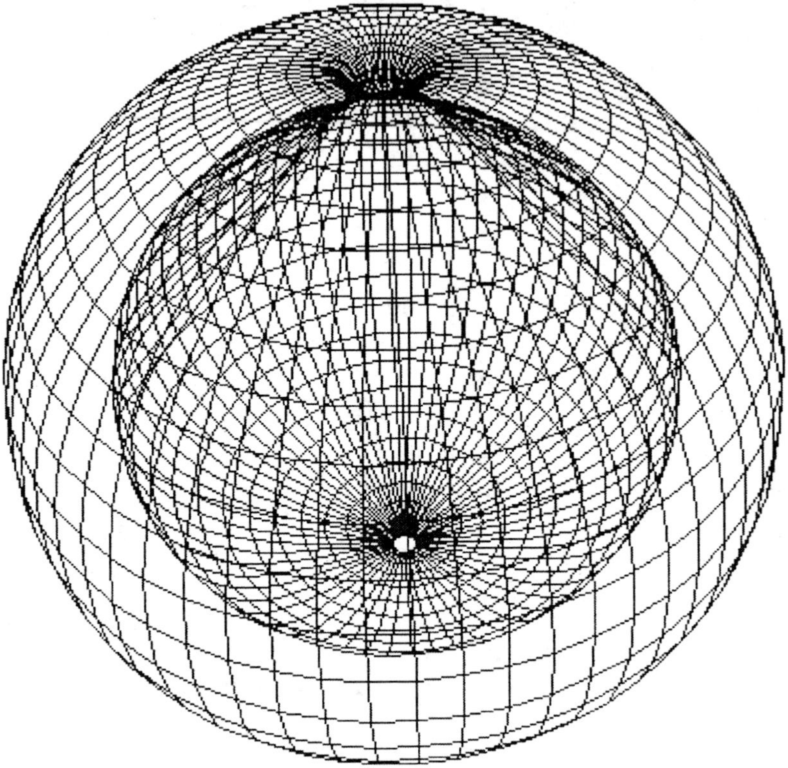

The Book of Enoch relates the plight of the one who has been cast out and down, in Chapter XXIX, verse 4:

> *And I threw him out from the height with his angels,*
> *and he was flying in the air continually above THE BOTTOMLESS.*

And earlier, in Chapter IX, verse 1, we have an actual geographic location as well as a description:

> *And those two men (angels) led me up on to the NORTHERN SIDE, and showed me there a very terrible place, and there were all manner of*

tortures in that place: cruel darkness and unillumined gloom, and there is no light there, but murky fire constantly flameth aloft, and there is a fiery river coming forth, and that whole place is everywhere fire, and everywhere there is frost and ice, thirst and shivering, while the bonds are very cruel, and the angels fearful and merciless, bearing angry weapons, merciless torture, and I said:

"Woe, woe, how very terrible is this place."

Here is the Biblical Sheol/Hell, as well as the tortures, magma rivers and ice rivers of Niflheim and Tartarus, and an actual description of the northern entrance to the underworld realm. These similarities are not superficial, but speak of a common link between all these traditions. The Hebrew Apocrypha also contain a detailed account of the misdeeds and trial of the NEFILIM, the "fallen angels" who interbred with human women, and were therefore incarcerated in caverns and prisons beneath the surface of the Earth; here is a strong similarity to the word for the Norse underworld of NIFLHEIM, which perhaps has it's roots in an even deeper antiquity than the Scandinavians themselves knew. Perhaps it is a reference to the subterranean abode, or the prison, of the Nephilim ("Those who fell from the heavens").....

It is the Book of Job, however, that truly lays the description out for all to see:

Job 26: 7-8:

He stretcheth out the north over the empty place ("void" in the Revised Standard Edition),

and hangeth the earth upon nothing.

He bindeth up the waters in his thick clouds;

and the cloud is not rent under them.

Job 26:10:

He has described a circle upon the face of the waters, at the boundary between light and darkness. (Revised Standard Edition)

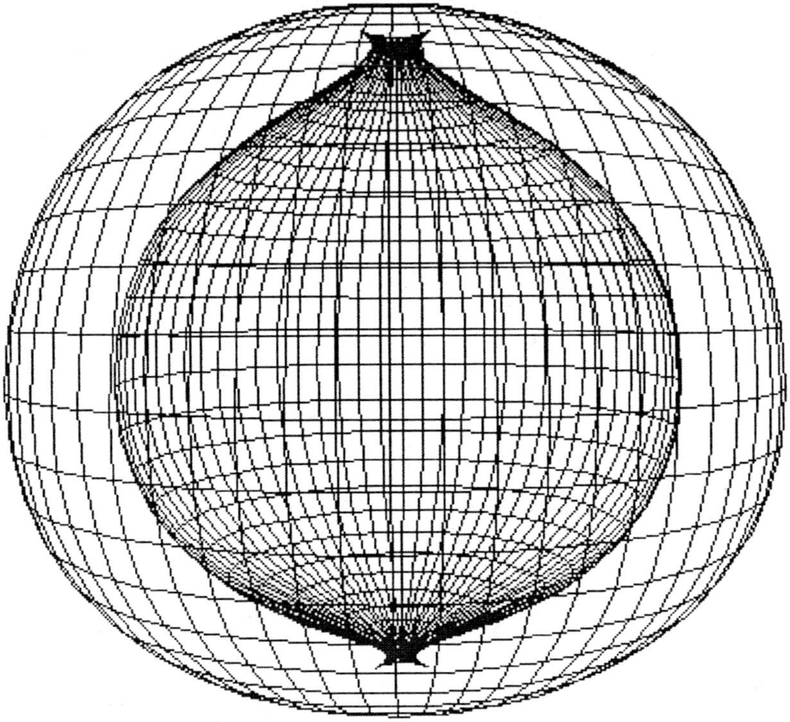

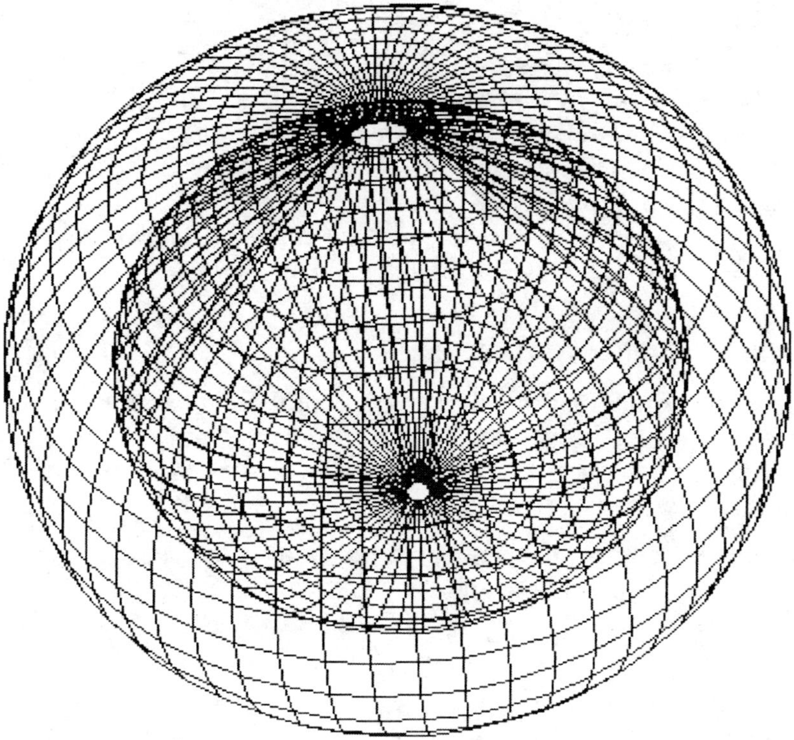

Job 38, verse 17, tells us:

> *Have the gates of death been opened unto thee? or hast thou seen the doors of the shadow of death (translated "gates of deep darkness," in the Revised Standard Edition)?*

Two verses later, in verse 19:

> *Where is the way where the light dwelleth? and as for darkness, where is the place thereof?*

And finally, the Arctic reference, from Job 38, verses 22-24 (Revised Standard Version):

> *Have you entered the storehouses of the snow,*
> *or have you seen the storehouses of the hail,*
> *which I have reserved for the time of trouble,*
> *for the day of battle and war?*
> *What is the way to the place where light is*
> *distributed, (meaning perhaps "equally dispersed and constant?")*
> *or where the east wind is scattered upon the earth?*

Could this last line mean that east and west essentially cease to exist at the ultimate northern reach of the globe?

These are some of the various scriptural, mythological, and folklore references to an abiding mystery; there are many, many more as well. The Roman sage Seneca, tutor to Nero and considered one of the most learned men of his day, stated:

"An age will come after many years when the Ocean (Oceanus) will loose the chain of things (binding ice-packs and icebergs?), and a huge land lie revealed; when Tiphys will disclose new worlds, and Thule no more be the Ultimate."

An amazing statement, to say the least; for what could exist further north than the "Land of the Ultimate North"--Ultima Thule--itself? Nothing on the surface of the world, perhaps.

What is the nature of the Void/Abyss, Ginnungagap and Tartarus, the Elysian Fields and the World-Tree/Igloo? What anomalous regions might lie hidden at the inhospitable polar regions of our planet, proof of a physical truth as unknown to most modern human beings, as the roundness of the earth was unknown to the philosophers and wise men of ancient and medieval times? And who are the cruel masters of this realm, the Jotuns, the "angels fearful and merciless," armed with "angry weapons?" Who is it that is "flying in the air continually above the bottomless--" and what is the nature, or means, of his flight?

(See Color Inserts)

Only those with true courage, true persistence, and absolute determination will find out. Today, the polar regions are closely monitored like no other region of the planet, by the most powerful governments of the Earth--but while imagery and data can be found for locations bordering the geographic pole (not the "magnetic" pole, which is an entirely different animal), almost nothing escapes into the public realm which shows the areas of the actual "poles" themselves -- particularly "photographic" satellite imagery. The most technologically-advanced nations of the world have divided Antarctica up like a frozen pie, and sit waiting there for--something. Will the wait end in a headlong rush forward into the unknown, or will it result in a panicked flight from an emerging Ragnarok? Cook and Peary, both honorable men but separate explorers, claimed to have first reached the point of the ultimate north--but perhaps their controversy, like the wonderings of so many others who "fell short" after feeling certain of the same success--is due to the fact that the inner rim of a cauldron is neither truly inside, nor out, but somewhere in-between. And sometimes, perhaps, when the "cloud is rent--" the clouds which almost perpetually cover the poles like a lid--one can look across at an opposite rim and see Admiral Byrd's "enchanted continent in the sky, land of everlasting mystery." Or see the Midgard Serpent, tail in his mouth, waiting for a release from an imprisoned tedium.

The northern hemisphere, showing the U.S. North of Cape Hatteras to San Diego; all of Canada, Greenland, Iceland, and the North Pole. Photographed from Apollo 11. Arrow point toward an unidentified object. NASA photo from Science Digest (August 1977)
(See Color Inserts)

Part IV: DEMONSEED: CHANGELINGS, HALFLINGS, AND HUMAN MONSTERS

Detail of a Sumerian underworld demoness, suckling her (hybrid?) offspring.

1. PREFACE

On the contrary, she seemed strangely proud of the dark, goatish-looking infant who.... was heard to mutter many curious prophecies about its unusual powers and tremendous future.

–H.P. Lovecraft, "The Dunwich Horror"

In his classic piece of horror fiction (later made into a mediocre film), "The Dunwich Horror" (1928), H.P. Lovecraft tells a tale of unspeakable terror set in a small New England town. Dunwich, Massachusetts, is an isolated hamlet which is decaying and filled with a sense of uncleanliness, where the inhabitants have interbred for some time, with one another, and eventually, with something else. Out of another "dimension," analogous to the underworlds or fairylands of old, a new source of genetic materal comes, when Old Man Whateley makes a deal with an underworld entity called Yog-Sothoth, and allows the demonic thing to mate with his daughter, Lavinia, impregnating her with a potential threat to the human species. Her son Wilbur Whateley, tall, hulking, and strange of shape through baggy clothing, is not all that he seems, and eventually his slain form is revealed:

> *Above the waist it was semi-anthropomorphic; though its chest, where the dog's rending paws still rested watchfully, had the leathery, reticulated hide of a crocodile or alligator. The back was piebald with yellow and black, and dimly suggested the squamous covering of certain snakes. Below the waist, though, it was the worst....*

The rest of the description of Wilbur Whateley goes beyond vertebrate characteristics, with mollusk and cephalopod features (aquatic or semi-aquatic?), limbs, and other members. The full horror of the description will not be repeated here in the interest of fair usage, but the story is recommended to any student or researcher of subterranean folklore. The resemblance of Wilbur to the Sumerian underworld demon called "PAZUZU," or to a variety of cryptid as well as "reptoid" forms from Ufology and folklore, is apparent. Also of interest is the fact that the story ends with the revelation that Wilbur had a twin, not quite as capable of masquerading as human as he had been...in fact, he is an invisible elemental giant of greatly destructive power. PAZUZU was of the underworld, yet was also a "demon of the wind," in effect paralleling both of the abominable brothers from "The Dunwich Horror".

An interesting piece of fiction, one might say, and indeed it is. But do things like this really happen in the real world? Do human beings, either willingly or unwillingly, make pacts or genetic links with nonhuman beings from beneath the Earth or elsewhere, contaminating the human gene pool even as they enrich the intrusive, parasitical one? Do history, folklore, and medical science hold clues as to this seeming fantasy being a reality?

Evidence for such activity would necessarily consist of two types: historical and folk accounts, and real persons of questionable and mysterious genetic heritage, a heritage which may result in unfortunate and sometimes inexplicable medical or physical conditions. As in the Lovecraft story, there is sometimes an element of the human family or gene-pool being interbred to some extent as well, and the extent to which they are interbred may have some affect on the likelihood of the non-human genetic characteristics being dominant upon occasion, even surfacing in future generations. This does not mean that these people are not human beings, although upon occasion this may have been debatable. It simply means that they have retained or inherited genetic characteristics from a contamination in the distant past, which is beyond their control or responsibility. As it turns out, both types of circumstantial evidence do exist, and in fact may indicate that the second type of evidence could be the result of the first, even if such trysts or cross-breeding took place decades or centuries before.

Before going further, it must be observed that unfortunate birth defects and malformations exist which doubtless do not fit into the evidentiary category described. No person should ever be shunned or otherwise mistreated due to any defect beyond their control. Negative behavior of this type is wrong in all instances, and should never be tolerated or condoned for any reason. Compassion is one of the fundamental aspects of true humanity, and many a true "monster" has held a beautiful or otherwise normal appearance, yet inwardly been a beast of the worst description.

To return to the study at hand, folklore and history are filled with tales of such half-human or genetically-contaminated unfortunates, and often the two anecdotal areas overlap. It must be observed that not all such liaisons have resulted in deformity or an otherwise evil condition or nature. How do such "alien" genes, possibly originating with pre-human races who shun the modern sun, enter the world of human beings?

There seem to be two methods: through forced interbreeding (abduction, rape, and the like), or else willing pacts, and even marriages. Closely-monitored or genetically-related groups seem to be the target of the latter, while the former appears more random at times.

2. IN DAYS OF YORE

'The Blatant Beast,' quoth he, 'I doe pursew,
and through the world incessantly doe chase,
Till I him overtake, or else subdew:
Yet know I not or how or in what place
To find him out, yet still I forward trace.'
'What is that Blattant Beast?' then he replide,
'It is a monster bred of hellish race,'
Then answered he, 'which often hath annoyd
Good knights and ladies true, and many else destroyd.
'Of Cerberus whilome he was begot,
And fell Chimera in her darkesome den,
Through fowle commixture of his filthy blot;
Where he was fostred long in Stygian fen,
Till he to perfect ripeness grew, and then
Into this wicked world he forth was sent,
To be the plague and scourge of wretched men:
Whom with vile tongue and venemous intent
He sore doth wound, and bite, and cruelly torment.'

–Edmund Spenser, *The Fairie Queene*, Book VI, Canto I, VII-VIII

Folklore is filled with tales about succubi or succubae, incubae, and demon lovers of various descriptions; these represent the negative, intrusive, and undesired type of encounter with such beings, whether they be visible (like "reptoids" and other Ufological types) or invisible (like Wilbur Whateley's brother, or his father, Yog-Sothoth). The other type of interbreeding is just as common in terms of anecdotal evidence, however. Sometimes fair, sometimes foul, the evidence exists, nevertheless.

If we look closely, we find that the wizard Merlin, of Arthurian legend and romance, was not the only person of note to have been rumored to have a "hybrid" parentage. Some accounts claim that his father was a sylph, or spirit of the air, but many other clues point to a "water-being," a merman, or something similar, for the "mer" in Merlin comes from the same root which gives us both mermaid, and the Merovingians, the dynasty supposedly descended from an aquatic demon or humanoid, according to their own tradition. This may explain in part Merlin's lust for Nimue, the Lady of the Lake.

The Isle of Skye is famous for just such a legend of the "willing" type. The tale goes that the Lord of Dunvergan Castle, Ian Ciar Macleod, entered a "fairy hill," sithean, or knoll (a cavern-world entrance), where he met and fell in love with a "fairy princess." One account states that she agreed to marry him and live in the surface world for twenty years. At the end of the twenty years, she left him (and. presumably, their offspring), and returned to her world, leaving behind a charmed "fairy flag" which the Clan Macleod still has, and displays, to this day. Another version states that, rather than a marriage, the situation was an affair, and after the birth of a son, the fairy lady took her infant and returned to her underworld home with a new genetic prize. The flag itself figures prominently in the lore and legends of the Macleod clan, with many stories and mysterious happenings associated with it. He is said to have received the flag, "Am Bratach Sith," in the 14th century, at a "fairy bridge" from whence his wife/lover departed. The bridge has significance, as both potentially being above or near an underworld entrance, and as allowing the fairy to "cross running water," something that many types of underworlders could not do. The latter myth is interesting, in that streams, brooks, and other fast-running waters often generate a strong magnetic current, which ties right in with the faerie aversion for some types of electromagnetism (as in iron). The castle Dunvergan and the bridge also sit on the banks of Loch Dunvergan, and lochs (lakes) are traditionally portals into cavern-realms or fairylands beneath their peaty beds.

Another version states that a Macleod went on a crusade to the Holy Land, where he met a seductive yet deadly female of demonic nature (a lamia or succubus), whom he bested in combat and drove back to the underworld–taking a scrap of her clothing as a memento when she fled. Generally, this legend is not considered the true origin of the flag, which supposedly has a reputation of good fortune and victory for the Macleods. Perhaps the crusader legend represents another encounter of a member of the clan with an underworld female, one less comely and pleasant of demeanor than the fairy lady in Scotland! Another legend states that whenever the clan was in need of help in battle and the flag was waved, armies of the "little people" would come swarming to the aid of their human kinsmen or in-laws. There are several other versions, or permutations, of the origin-story, and additional tales of the Fairy Flag of the Mcleods.

This is reminiscent of the famous Scot, Thomas the Rhymer, who was said to have been taken away for a seven-year love affair by the subterranean "Queen of the Fairies."

Encounters with such "elemental" beings are not all so so charming. Elementals are notoriously capricious and unpredictable, and are usually indifferent at best, and dangerous at their worst. Theosophists, Rosicrucians, and Spiritualists tended to theorize that "fairies" were elementals, but were also spirits (existing on "another plane"); yet the Neoplatonists, and later the alchemists of the middle ages, believed them to be of a finer and more subtle type of matter, somewhere "between man and the angels." They had flesh, after a fashion, and fleshly desires to go with it. The wizard Merlin (his name meaning "of the sea" or "of the water") was supposedly the result of the lust of an elemental for a human woman. While there is traditionally some inter-relationship between the four types of elementals (sylphs, undines, salamanders, and gnomes), it seems that many "fairy" types of beings from Northern and Western Europe have a close relationship to the gnome (underworld) and undine (mermaid/merman) varieties. According to Reverend Robert Kirk (*The Secret Commonwealth of Elves, Fauns, and Fairies*, 1691), the fairies who abducted and later released him were composed of "congealed air," or a finer type of matter. Yet, in defense of the spiritual aspect or theory, fairy associations and fairyland entrances were often with subterranean regions or necropoleis, such as barrows, tumuli, ruined hill-forts, mounds, and the like. Caves and wild gorges and chasms also had the same types of fairy associations, however, as did crevices or cracks in rocks which were too narrow for human beings to enter.

Through the centuries, some have theorized that the fairies of the British Isles were fallen angels, the souls of the pagan or unbaptized dead, or the actual hidden descendants of races (or species?) of man which had been defeated by the Celts and driven into hiding, eventually beneath the ground. Here, after long ages, they often became diminutive in form and stature, due to cramped quarters, poor nutrition, and inbreeding. Walter Evans-Wentz reports an interesting belief (as expressed in *The Fairy Faith in Celtic Countries*, 1911, pp. 166-67), held by David MacRitchie previously, who in his book *The Testimony of Tradition* (1890), noted the similarity between the word *pixies,* or *pixys,* and *Picts,* a Brone-Age, pre-

Celtic people whom he saw as non-Aryan (a possibility, at least before the coming of the Celts) and dwarf-like, the latter now known to be false. But what might years of interbreeding in a subterranean environment do to the stature of a people, or isolated groups of survivors in hiding from taller, stronger invaders? The finding of apparent Neolithic dwarf remains–an entire tribe of dwarfs–in Switzerland in 1893, lent credence to the theory of ancient, "pygmy" races who had been driven into a slinking, clandestine, and eventually subterranean existence throughout Europe. To the Victorians, this led to the conclusion that dwarfism and midgetism were possibly due to the "racial residue" of Cro-Magnon types interbreeding with these ancient peoples, and that the fairies were simply "natural" and "earlier" forms of man.

For a time, even the Spiritualists and Theosophists adopted this theory as their own. This didn't last, and the soon these groups were back into the esoteric, "elemental" or "devic" (nature spirit) concepts.

An unpleasant account and tradition come from another Scottish castle and family, and spill over into recent history.

North of Tayside is Glamis Castle, in Angus, ancestral home of the Bowes-Lyons or Earls of Strathmore, as well as the Queen Mother (Lady Elizabeth Bowes-Lyon) and hence, the current Windsor Royals. It's also the birthplace of Princess Margaret. Glamis (pronouned "glahms") is surrounded by a variety of eerie and disturbing legends, ranging from many bloody, ghostly hauntings (including by the historical Macbeth, and a tongueless woman with a bleeding mouth), a female vampire who shows up from time to time, and a strange creature with thin arms and legs and a beast-like head, described by one percipient as "a hellish combination of all things foul and animal." This young lady, sleeping in a "haunted tower room" in the castle in the 19th century, was awakened by the strange apparition or creature, which vanished when someone else rattled the bolt on the door before entering the room. Was an underworld or non-human rapist interrupted or frightened away?

But these are not the greatest terrors of Glamis, which has a horrific secret tradition. For centuries, there have been rumors of the periodic births of "half-human" or otherwise deformed "monsters" at Glamis Castle, and their imprisonment, for life, in some of the secret chambers which allegedly exist in the structure. The most recent such birth is said to have occurred in 1800, and the deformed heir to the Bowes-Lyon line was described as having "a huge, egg-like body" (shades of Humpty Dumpty!) and stunted or stubby limbs. He has also been described as being "neckless," and extremely hirsute or hairy. Others have insisted that he was merely mongoloid (suffering from Down's Syndrome). The tradition also maintains that he was still immensely strong and inordinately evil, and had to be locked away from the rest of the family in secret, for both his safety and theirs. He allegedly did not die until the year 1921, meaning that he lived for 120 years in nearly-complete secrecy! This negates any rumors of Downs Syndrome, as those who are stricken with this unfortunate condition have generally not been known to last into old age. Supposedly, during the hidden heir's lifetime, each successive Earl of Strathmore was told of his secret existence when they reached the age of twenty-one, and shown the "rightful" heir.

One tale has it that the first Lord of Glamis, after gambling with the devil (or an underworld being of some sort) at cards on the Sabbath, brought down a curse upon future generations of his line. Could it be that the stakes of the card-game were actually more than just his soul, and that he and his were cursed to make payment in genetic exchange, down through the centuries? Could it be that the genetics of families which tend to be somewhat interbred or interrelated, whether isolated by geography, or of "noble" lineage, are nurtured by unseen manipulators for certain desired genetic traits?

There is a family secret surrounding tales of the Glamis "monster," which is said to be told only to each male heir of the Bowes-Lyon family when he attains majority (reaches the age of 21). At one time this supposedly dealt with keeping the deformed, rightful heir locked away and his existence concealed, even from his own family (supposedly, only four men at a time were allowed access to his cell, and knew of his existence, one of these being the family "factor" or steward, and one the attorney). This was first told by the second Lord of Halifax, who heard it from the 13th Earl of Strathmore himself.

An intrusive, lustful visitor.
From the Violet Fairy Book, by Andrew Lang.

This "dread ancestral secret" may have also dealt with the family tradition, or curse, of having an occasional birth of a less- or other-than-human child born to the clan. There is said to also be a secret concerning a "vampire-like" or blood-lusting being which is periodically born within the family.

All of this raises the obvious question: What is the source of this perhaps-repeating genetic anomaly? Could it have anything to do with strange nocturnal things, male or female, which creep into sleepers' rooms by night? Or was there, in the distant or recent past, a willing liaison now and again, with the "unseen rulers of this world," mentioned in the Book of Ephesians, representatives of the gnome, troll, fairy, elemental or reptilian races which preceded man on this planet?

It should be noted that upon several occasions, different 'factors' of the family, trapped at the castle by either snow-storm or other inclement weather, have refused to stay on the grounds, out of apparent terror based upon the secret knowledge to which they were privy. Another story (All the Year Round Magazine, 1880) states that, during the lifetime of the monstrous heir, a workman accidentally discovered his chamber, and was "persuaded to immigrate to Australia (leave the country and keep his mouth shut?)." Another version states that a stonemason accidentally broke into the chamber, and "died of shock" (did he actually meet his end at the hands of whatever was in the room?). His wife was paid off and packed off to Australia so that she wouldn't ask questions.

As for the unfortunate, deformed and hidden Earl...He is rumored to have been sealed inside his secret room after death, where his remains rest in secret to this day, his restless spirit joining the many others which have been reported by witnesses at Glamis, through the centuries. To add another clue to the mystery, and credence to the periodic 'strange births,' there is said to be a family-type portrait in the drawing-room, which depicts an earlier Earl of Strathmore, his two sons, and a horribly deformed and unsightly dwarf whom no one can (or will?) identify....

Additional legends about Glamis involve a giant ghost (thought to be the ghost of the card-playing first Earl, "Beardie"), a ghost with "long, hairy arms," and the local folk belief that "monsters" are still born into the family upon occasion, or there is still one, hidden on the premises, who has never died.

3. UNHOLY LIAISONS

But full of fyre and greedy hardiment,
The youthful knight could not for ought be staide;
But forth unto the darksom hole he went,
And looked in: his glistring armor made
A litle glooming light, much like a shade,
By which he saw the ugly monster plaine,
Halfe like a serpent horribly displaide,
But th' other halfe did womans shape retaine,
Most loathsome, filthie, foule, and full of vile disdaine.

–Edmund Spenser, *The Fairie Queene*, Book I, Canto I

He rose to his feet and backed away, and as he did, something floated up out of the well. What it was, Conan did not know. He could see nothing in the darkness, but he distinctly felt a presence–an invisible, intangible intelligence which hovered malignly near him.

–Robert E. Howard, "The Scarlet Citadel"

Who or what might come to men and women in the dead of night, for purposes of seduction and genetic exchange? The classical lamia is the same as the succubus, whose male counterpart is the incubus. To the Irish, the female of this species or type was called *Leanan-Sidhe* (lan-awn-shee), and on the Isle of Man, she was *Lhiannon-Shee*. Both variants sought the genetic influx of a human male, and in exchange gave him fame, creative inspiration, or other abilities (much like the ancient Greek Muses). Invariably, such a relationship would end with the premature death or "wasting away"–sometimes attributed to vampirism–of the man. The Manx minx also had a close association with deep wells, from which she would emerge from her underworld home in order to seduce a lover, or enter into an agreement with him.

This latter type has a close parallel in one aspect of Tantric Yoga and in Hinduism. The "Left-Handed Tantra," or black magic, and its adepts, have knowledge of and supposedly interaction with a nearly identical being. This female entity, called Mohini, comes into the night-time world by means of ascending through a deep well. She is enchantingly beautiful, yet simultaneously horrific, unearthly, and deadly. Like her British Isles counterparts, she also seeks a male lover and human genetic material, presumably for the purpose of bearing a hybrid child.

In "Secrets of the Left-Handed Tantra," A.T. Das relates his own account of an encounter with a subterranean demoness who emerged from a stone-lined well (revealing a link both with the underworld, and with underground water-sources, much like Celtic fairies with their wells and "fairy bridges"). After a terrifying encounter in which he and a friend narrowly escaped her clutches, the experiencer was told:

> *'"You saw Mohini, a demoness from the underworld. Had you known how, you could have entered a pact with her for the next cycle of Jupiter (twelve years). You promise to satisfy her lust once a month, and she will do your bidding in return–protect your property, destroy your enemies, whatever.*
>
> *"But a pact with Mohini is very dangerous. When she comes for sexual satisfaction, she may assume eighteen forms in the course of the night, expecting you to fulfill the demands of each one. If you cannot, it will cost you your life. And if during the twelve years of your relationship with her you have an attraction to another woman, that will also cost you your life. You suddenly vomit blood - finished."*
>
> *I asked, "Why was she attracted to the white stones?"*
>
> *"Mohini draws energy from the male sexual fluid," he answered. "Besides the pleasure of sex, this is her main interest. Of the bodily fluids, saliva is the most similar to semen; that's why throwing a white stone upon which you've spat is a sure way to divert her attention. People who drool while sleeping unknowingly invite this kind of succubus to take control of their bodies."*

In this tradition (or reality) we see a parallel to the age-old fable of Faust and others, "making a deal with the devil;" here is Tannhauser, and his temptation by an underworld queen. In the *Mahabharata*, Arjuna begets a half-human son due to the seduction of the nagini, Ulupi. In the original Hindu tradition, Mohini was a female form utilized by Lord Shiva in order to seduce, confuse, and defeat the evil asuras and others; the tantric variant of Mohini, as defined in this account, is not the original but is apparently a type of *nagini*, or female naga (serpent-person). In her own realm, she is partially reptilian, and is usually represented as being a snake or crocodile from the waist down. When coming to the surface for purposes of seduction, however, she creates an illusion of a fully human form in order to attract a human mate. She "glides" rather than walks, and although the percipients believed that they saw her legs, it is clear that another type of locomotion–possibly serpentine motion, or levitation–is in use. This use of illusion, or "glamour" (the same type of power attributed to Western fairies), is indicated by the following passage from Das' account:

> *"Ten minutes after my friend fell asleep, I trembled, as a cold tingle crept up my spine. Leaping to my feet and turning around I saw something that made my heart almost stop. Bathed in the moonshine a tall statuesque women stood on the well's rim. I was sure she didn't walk to that place because I was watching. Her eyes were closed. For a moment I wondered if she was a sleepwalker. In face and physique she did not resemble an Indian woman . She had long loose hair that hung down over the front of her body to her knees - otherwise, she was naked. She was hauntingly voluptuous in a way that was both enticing and frightening.*
>
> *Staring open mouthed at this apparition I nudged my sleeping friend with my foot. He sat up with a shock, turned to see what I was looking at, then gasped and scrambled to his feet. At once her eyelids lifted revealing twin orbs from hell to penetrate the darkness with a glare like the eyes of a tigress. She fixed those terrible eyes upon mine and stepped off the well alighting to earth as if she was not heavier than a wisp of cotton. The woman's legs propelled her forward. I cannot say she walked or ran or floated for these words will not simply be able to give you an accurate picture as how she advanced upon us. Her legs moved without*

bending at the knees making swift little steps of such fluid effortlessness that I was reminded of the locomotion of a centipede. It was almost as if below the waist her body was motorized for when her legs started her head and upper torso with her limbs snapped back slightly from the sudden forward motion at least to our vision."

Naga and Nagini.

Picture a cobra's lower body with a woman's upper form, and the visual effect of forward motion might very well be the same.

As of late, Conspiracy and UFO literature alike abound with similar accounts, the only difference being that most of the reptilian seducers or rapists are male, and the human victims are female. Often, these accounts also involve abduction to underground bases or caverns, cavern-worlds, or close proximity to a cave or cavern in which the rape takes place.

Another version of the succubus or incubus attack or rape occurs when the percipient awakens fully (at least mentally), can see and feel, but can't move a muscle. Medical Science refers to all such cases as "sleep paralysis," and states that the condition is due to an "awakened mental state, while the body is still asleep," or the person is in a "hypnogogic" state. While this may be true in some cases, there are those who maintain that this is not the case in their particular experiences. Often these incidents are accompanied by "waking dreams" or "hallucinations," in which the victim feels, senses, or sees someone or something else in the room or beside the bed, or even feels a bodily or electrical pressure or weight descend upon them. A few female victims have reported actual physical intercourse with an unseen entity or group of entities, and this has, according to them, led to "rape" by invisible presences at later times, when alone and fully awake.

Some see a psychological connection or cause to the latter type of supposed hallucination, often arising, in theory, from childhood trauma or molestation. But not all experiencers have such a thing in their chest of childhood memories or secrets. Some victims have noted that the emotion of fear increases the severity and frequency of the attacks, while praying to God for deliverance halts them, or ends the phenomenon completely. Rather than a hallucination, this would seem to indicate a conflict with the "unseen," operating on the terms of the unseen, or spiritual, world.

This calls into question the nature of this intrusive reality, and its intrusive, abusive inhabitants. Some evidence indicates a "physical reality" which needs or requires blood, seminal matter, sperm and ova, and perhaps an ongoing ritualistic pattern of inflicting pain, fear, desire, and even servitude, for purposes of emotional or energy vampirism and behavioral conditioning. Folktales and mystical traditions are filled with accounts of these beings, and their ability to "shape-shift," as well as alter their forms, from seemingly fleshly, to barely substantial or opaque, to absolutely transparent or invisible. If it is possible for spirit to become physical, for beings from some other dimension or ethereal reality which infringes upon this one to occasionally manifest here for reproductive or other predacious purposes, this would explain much, including the fairy/spirit aversion to cold iron and electromagnetism, which might be due to a field disruption that could banish a 'temporary form' from our reality on the spot; and to the aversion to direct solar radiation, which might have a similar effect on a temporary living physical body or vehicle. The Apostle Paul may have been alluding to this in Ephesians 6:12, when he named a hierarchy of enemies who oppress and victimize mankind, and oppose the will of their:

> *For we wrestle not against flesh and blood, but against principalities [fallen angelic types and kingdoms], against powers [fallen angels], against the rulers of the darkness of this world [the unseen predators under discussion here, as well as possibly those human beings who interact with or serve them], and against spiritual wickedness in high places [the etheric and ufological connection, which has been often noted by researchers].*

Recently an actress with a regular role on a hit television series revealed in an interview in TV Guide Magazine (Sept. 14, 1999), that she had been visited by just such an entity:

> *I had a dream a long time ago. It was so real. ... It wasn't even a dream. I was sleeping on my futon on the floor, and some sort of spirit came down from God knows where and made love to me. It was sheer bliss. I felt everything. I climaxed. And then he floated away.... I've never had such an intimate relationship. I've never said anything to anyone about it until now, because I thought no one would ever believe me ... I feel like the spirit follows me everywhere. Isn't that weird?*

"Weird" may not be as appropriate as "sinister," insomuch as she, and her descendants for centuries, could be plagued and in all likelihood terrorized by this type of continuous intrusion. What would have happened to her had she resisted? Or, how long until she experiences the forced brutality that other women have received? The truly frightening part of this scenario is how willingly she submitted, and even seemed to welcome, such treatment. Does a life-time of psychological conditioning or brainwashing come into play?

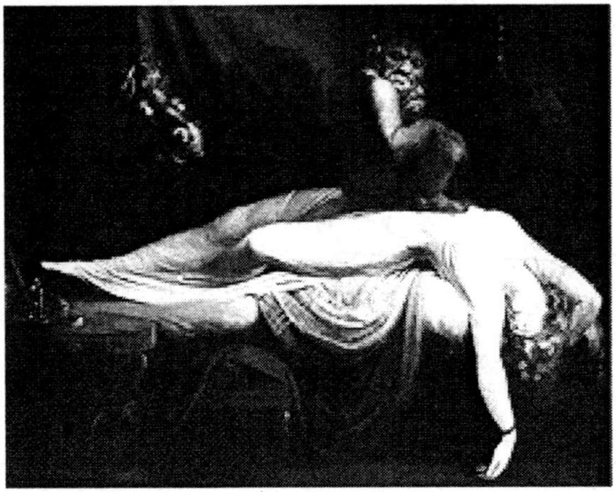

Two views of the incubus or hag-ridden phenomenon, by Fuseli.

(See Color Inserts)

Modern-day accounts of "sleep paralysis," "alien abduction," and rape scenarios would indicate that age-old patterns of victimization–whether the perpetrators are flesh, spirit, or both–are still ongoing on the Earth today. Do other contemporary accounts exist? Sean K., a well-adjusted young man who has had several experiences that he is seeking to understand, had this to say:

> *I have personally had hypnogogic experiences and lucid dreams with disturbing content as a result of spontaneous awareness during sleep paralysis. I've experienced this on countless occasions since I was sixteen years old.*
>
> *One one occasion I felt like I woke up in bed laying on my side and opened my eyes. I felt and saw thin red leathery batlike wings draped over me, such as the posture of a vampire from a Hollywood movie while*

preparing to bite its victim. I instinctively fought it off and it fled without a fight. Then I remained "stuck" in this aware yet not fully conscious state until an absorbing energy that feels and sounds like the buzzing of a bee hive came over me, then I woke up to full consciousness. I have on occasion also seen a classic "devil" in one of these dark lucid dream states. Considering my lack of religious background, these literal translations of devils and dragons would seem unlikely products of my imagination but there they were. Just my thoughts and opinions.

These types of experiences are not always accompanied by visual input, but the incapacitating effects remain the same. Sean further related:

> *On Halloween night of this past year I had an experience that shook me up badly. It was the first time I was so close to being fully conscious when the "entity" and paralysis struck. It struck terror to a level few of us experience. I was reduced to the state of a babbling child. In fact it made me feel exactly as I did when I was a crying three year old.*
>
> *There were no visions or creatures involved, just the presence, vibrational force/energy and paralysis but the terrifying aspects of this are how deeply the entity can penetrate one's "soul", how it can violate ones inner self to the degree it can. I mean we all think of ourselves as US and as sacred in a sense. No one can take up our space but us. Well, not so with this force. It violated me on a very personal inner level and also read my thoughts. In fact I felt something reading my thoughts seconds before it struck. I don't like the goose bumps I'm getting even typing this.*
>
> *I am going on 39 years old so having experienced this phenomenon since I was 16 you can imagine how many stories I can dig up.*

Sean is not alone in his experiences. Other men and women have related similar accounts, ranging from being held immobile, with or without of glimpse of the entity responsible, to moving from an erotic dream situation of extreme realness and into a similar hypnogogic state...sometimes glimpsing that which is inducing the sensations of the dream which had been underway. These intruders range from amorphous forms or beings of light or shadow, to "alien hybrids," "reptoids," "old hags," "devils," or alluring men and women. All are apparently different MASKS for the same phenomenon.

"Night terrors," "old hag," incubus, succubus, "reptoid," or demonic... all faces of the same phenomenon, throughout history.

Many victims of this force, or these beings, have come to realize that there is a connection between experimenting with occult or transcendental techniques, and the sudden beginning of the types of attacks described. These types of activities often seem to trigger an onslaught of the phenomenon, and often this can only be corrected by serious efforts of what can only be called "a return to religion."

According to the accounts of witnesses of another type of event, there is a definite connection between some types of physical "cryptids" (such as Mothman and El Chupacabras), and the supernatural realm. But this does not negate their physicality, or the very real possibility of their origin in an alternative biosphere, which may be concealed within our own planet. The Nephilim (Annunaki) in particular, banished beneath the Earth's surface, were said in the Apocryphal Books to have performed a wide range of genetic experiments, combining different types of creatures and men, and creating many chimaerical forms (like El Chupacabras). The description of

Mothman matches the bas-relief carvings of ancient Sumer and Babylon of UTUKKU, guardians and errand-runners of the underworld, who were humanoid yet reptilian and bird like in form.

Prolonged study reveals that there are paranormal aspects to all or most of these phenomena. The "reptilian" aspect of many of these beings, even the spiritual or supernatural ones, may have more to do with their very ancient (pre-mammalian) time and place of origin, as well as with their ability to present "glamour" or illusion when they appear briefly in our surface world. The tantric account is a perfect example of this. The description of the locomotion, distinctly reptilian or serpentine, of the female entity holds many clues as to the origin of that being, whether physical, "spirit," or something of both realms.... "She" was expressly called a "demoness" by the yogi who sent her to the witnesses.

In the west, demons are often wrongly thought of as "fallen angels," which they are not. They represent the restless spirits of what the Torah calls "the people of old time," destroyed during the rebellion of their fallen angelic masters yet still enslaved to the same masters, their previous civilization "thrust down" beneath the ground:

> *20 When I shall bring thee down with them that descend into the pit, with the PEOPLE OF OLD TIME, and shall set thee in THE LOW PARTS OF THE EARTH, in places desolate of old, with them that go down to the pit, that thou be not inhabited; and I shall set glory in the land of the living;*
>
> —*Ezekiel 26:20*

These are the equivalent of the asuras and rakshasas of Hinduism, and perhaps to some forms of fairy-type beings. There is also a very strong relationship or kinship between them and the "extant" or physical forms, such as some so-called cryptids, or unknown humanoids, which circumstantial and anecdotal evidence would indicate are the descendants of the same ancient races or specie which are "demons" in the spirit world. They are interrelated, representing two aspects of one unseen kingdom or force that has haunted and plagued mankind for thousands of years.

Lilith, mother of monsters, seductive demons, and abductor of babes.

Note the wings, similar to those of the UTUKKU.

For instance, in the Torah/Bible, there are mentions of those physical beings in the wastelands and ruined cities, which are translated as "satyrs" in the King James version, but that represent a type of "hairy humanoid" that today we might call Sasquatch, or maybe even El Chupacabras! In the same verses, there are mentions of Lilith herself, but her name is mistranslated as "owl" in the King James translation (the "owl" was one of the animals associated with Lilith), and as "evil spirit" or something similar in other versions. These type of phenomena—demonic (spiritual) and "monsters"—go hand-in-hand, which may account for some of the confusion of types and nature from time to time.

In fact, while Lilith was a "woman" of some pre-Adamic sort, she is also considered to also be an evil spirit, and in Hebrew/Babylonian traditions, is the mother of all succubae and incubae herself, or their spiritual progenitor. From ancient times to the present, esoteric tradition has held that she visits "those men who sleep alone," seduces them through their dreams for purposes of extracting semen, or else assaults them in the form of a "night terror," "old hag," or "sleep paralysis" episodes for the same purpose. From the semen collected, she is said to engender a partially physical, partially ethereal brood or race, called the "Lilim" (Babylonian-Assyrian lilu/incubus and lilitu/succubus). These beings can be visible or invisible at will, physical or nebulous, and act out the same evils as their ancestress or mother. They are also the same entities that are sometimes glimpsed by the victims, often with a reptilian appearance, and sometimes with other forms, which can range from satyr or "imp-like" (the modern "gray alien abductor" similarity is apparent here), to attractive, hideous, hag-like, or horrid. One of the only defenses against the whole family, in ancient times at any rate (more effective spiritual protections have come about in the ensuing centuries), were amulets bearing the names of the three angels who, according to Hebraic tradition, punished Lilith for breeding with the demons of the underworld. Their names were Senoi, Sansenoi, and Samangeloph.

Another underworld-originating, vampiric succubus of Assyro-Babylonian culture was Lamastu, who preyed particularly on women in labor and on sleeping human infants (the latter habit was also a favorite of Lilith herself). Like many underworld beings of the Ancient Near East, she was depicted as having a canine or leonine head, other animal and reptilian characteristics, and a humanoid form. An ancient rivalry existed between Lamastu and Pazuzu, and amulets of the male demon-beast were believed by the ancients to ward her away, as Pazuzu had to routinely come forth from the subterranean realm and drive Lamastu back beneath the earth. The Greek equivalent to Lamastu, Lilith, and the lilim, were the mormo(i), the lamia(i), and the empusa(i). Like the Hindu nagini, the lamia was often depicted as half woman and half serpent. The empusai were believed to be the daughters of Hecate, the goddess of witches who lived in the underworld and also represented a variant on the Lilith theme. Her shape-shifting daughters would emerge from the underworld in order to seduce and obtain genetic influx from mortal men, before killing them. All of the various entities of incubus or succubus type which have been mentioned were believed to be vampiric, as well. The empusa also has a Highland Scottish counterpart in the baobhan sith (baavan shee), a shapeshifting seductress and man-killing vampire (and of course, also considered to be a "fairy"). Another Scots/Celtic variant is the Glaistig, essentially the same except that she hides the legs of a goat or hooves beneath a long, green dress. And then there was the Teutonic, Scandinavian, and Anglo-Saxon Mara (Slavic *Mora*), the succubus whose name gives us the word *Nightmare*....

On a related note, it should be suspected that many of the "shape-shifting reptilians" which many witnesses report are actually representative of an accidental glimpse of a demonic entity or power which has attached itself to the person in question, and which has such a strong presence or personality that it somehow manifests to view—more to the "psychic vision" or stimulation of the visual cortex of the viewer, than in our physical world or native dimension. This may be due to the same type of "seeing" which is accomplished by the succubus or incubus victim when they "see" an intruder or attacker, an altered state of brainwave activity (hypngogic, or hypnopompic) which might be permitting a greater sensitivity of the visual cortex to subtle magnetic fields or presences. Upon occasion, a lucky camera or video-feed might pick up such a manifestation, but it is generally put down to technical difficulties, equipment failure, and so forth. And these types of incidents, in and of themselves, could be seen as potentially deceptive or manipulative.

There is a definite need and thirst for blood and sacrifice among those of this mostly-unseen kingdom. As postulated previously, some types of creatures which take blood and the "sacrificial organs," particularly the livers, from pets, livestock and even human mutilation victims, may be on a mission to obtain these substances for their unseen, mostly amorphous or invisible (spirit or other-dimensional) masters in the traditional underworld, and/or another dimension. This would explain much in regard to the mutilation and exsanguination phenomena. By the same token, the incursions and predations of the incubus, succubus, "fairy lover," or "reptoid rapist" might be for the purpose of obtaining viable DNA and reproductive materials. In medieval esoteric tradition, the succubus was said to "carry" the seed of her victim to the underworld, where it would be put to use by the diabolical forces there.

Another interesting aspect, as discussed at length in "The Deep Dwellers", as well as in "Hidden Neighbors", is that the ancient Sumerian underworld entities called "GALATUR" and "UTUKKU" were in the service of the unseen "underworld gods," and as part of such service, were required to fetch home the fruit of the sacrifices and burnt offerings, which generally consisted of the blood, liver, and other "sweetmeats" of the sacrificed animal. The GALATUR (who were also in charge of abducting surface-world humans, much like alien abductors do today) had a distinct resemblance, in their description, to both the gray aliens and some descriptions of El Chupacabras; the UTUKKU resembled Mothman, El Chupacabras, and even some descriptions of the Jersey Devil, in many respects. The underworld gods, and their servants as well, were often portrayed as being reptilian in some feature, or even in their entire forms, as in the case of the demon-servant called PAZUZU. This being matches descriptions of modern-day ufological "reptilian humanoids" or "reptoids," almost exactly.

All of these things are nothing new. The same forms—physical, spiritual, and archetypal—repeat over and over again through the centuries....

4. THE BEAST WITHIN

I have heard relate that such a pair

they have sometimes seen;
march-stalkers mighty the moorlands haunting,
wandering spirits; one of them seemed,
of womankind; the other, accursed,
in man's guise trod the misery-track
of exile, though huger than human bulk.
Grendel in days long gone they named him,
the folk of the land; his father they knew not,
nor any brood that was born to him
of treacherous spirits. Untrod is their home;
by wolf-cliffs dwell they and windy headlands,
fenways fearful, where flows the stream
from mountains gliding to gloom of the rocks,
where lies the underground flood.

—BEOWULF

So Conan, glaring from under his tousled mane, saw the figure of Natala writhing in the lustful grasp of a black nightmare shape that could only have been bred in the lost pits of hell.

—Robert E. Howard, "The Slithering Shadow"

As for Tsotha—Men say that a dancing-girl of Shadizar slept too near the pre-human ruins on Dagoth Hill and woke in the grip of a black demon; from that unholy union was spawned an accursed hybrid men call Tsotha-lanti—

—Robert E. Howard, "The Scarlet Citadel"

Savannah, Georgia, is renowned in some circles for its rich tradition of haunts and spooks, mysterious lights and figures, and a history which is in general rich in terms of paranormal occurrences. Yet none are so spine-tingling as the story of a supposed French-Huguenot immigrant, Rene Ash Randolier. His true origin, or parentage, is unknown.

Rene Randolier, born in 1777, immigrated to Savannah, where he took up residence in a part of town which was the haunt of prostitutes, sailors, and the poverty-stricken. He was generally avoided or even shunned, for his appearance was not exactly human—he was of a height far above the norm, and his head, face, and body were covered in an animal-like coat of hair. Black slaves in particular held him in great awe or apprehension.

He was at times observed capturing and torturing pet cats and dogs, and was also seen in the act of dismembering chickens and goats, as many a "hairy humanoid" and other unknown creature like El Chupacabras has been said to do. These weird, almost sacrificial activities were tolerated by the neighborhood at large, until he was implicated in the murders of a woman and two little girls, sometime after 1796. As he had done previously, Rene fled into the extensive system of tunnels and catacombs which lay beneath Colonial Park Cemetery, by some accounts dragging the bodies of the little girls with him. He had previously lurked in this huge, walled graveyard, allegedly digging up graves in ghoulish (ghulish) fashion, hiding out in the catacombs beneath, and generally having such a run of the place that people had avoided it for fear of him.

He was finally hunted down by a mob, and hanged by the neck for his various crimes. Local legend maintains that his restless spirit haunts the area of his old stomping-grounds, and particularly the cemetery and the tunnels under it, to this day. According to some, these tunnels contain evidence of graves being dug into from underneath, and the bones of the occupants having been split and gnawed upon.

Little to no information exists on the true origin and lineage of Rene. Did he represent some sort of atavistic form of man, a "throw-back" to an earlier hominid type? Or was he, at least in part, something else, something other than human? The ghuls of Arabian folklore are hairy, cannibalistic eaters of human carrion, and mostly subterranean in terms of their housing preferences. Exactly what was Rene Randolier?

Neanderthal sites have been uncovered, always in ancient caves, which reveal butchered, split, and gnawed-upon bones of their own kind...and other ancient, pre-Adamic hominids may have engaged in similar habits. Going back to the most ancient accounts, we find that there are hints of links or relationships between earlier forms of "man" and their reptilian-demonic precursors, as in the case of the living, breathing and procreating Lilith, who sought refuge, and interbreeding, with such underworld beings after being outcast from the presence of God and Adam. This is echoed throughout myth and literature, with Caliban and Ariel, or Grendel and his unknown, demonic father, linked in an unholy kinship. This may be interpreted literally, or it may represent a schism between two somewhat-related but very different species, which occurred in the very distant past and which haunts the world to this day. Are many of the more outrageous and mysterious unknown creatures of the world, particularly "hairy humanoids," somehow related to this primeval rivalry between hominid species?

What was the cause of Rene's brutal activities, his cannibalistic and ghoulish feeding habits? Why does a somewhat subterranean connection also exist? It is recalled that the ghuls and nagas are said to eat the human beings of the surface world, upon occasion; Richard Shaver made it quite clear that his "deros," in addition to being sexual peverts and torturers, were also cannibals or eaters of the flesh of the human beings on the surface. Is Rene the only "degenerate cannibal" with an urge to literally "go underground" for refuge? Hardly.

A century before Rene walked the Earth, or scuttled beneath it, another infamous band of miscreants engaged in similar activities in the south of Scotland. Sawney Beane was born near Edinburgh, during the reign of James VI of Scotland, (James I of England). Sawney was a mean-spirited and violent sort, and to avoid honestly working for a living on the family farm, he ran off to seek his fortune, taking his girlfriend along as help-mate. The two of them soon set up housekeeping in a deep cave which was situated on the Scottish coast, near Galloway. This cave is said to penetrate more than a mile into the ground, with many twisting passageways, and unexplored tunnels leading off into Stygian depths.

For some time the pair made their living as robbers who murdered all their victims, and then, money being short among travelers and meals being few and far between (or so the story goes), they decided that they didn't need to buy meat after all. They began to pickle, salt-cure, and otherwise subsist off of the flesh of their unfortunate human prey.

At length, Sawney and Mrs. Bean starting procreating like rats, producing fourteen children in all, who in turn gave rise to a second generation of eight grandsons and fourteen granddaughters, by incestuous means. This inbred crew held the countryside in terror for over two decades, murdering travelers (who would seemingly "vanish") on lonely roads by day or night, yet members of the Bean clan were never suspected or apprehended when they ventured into towns and hamlets. And this entire brood of troglodytes called the deep, seaside cave their home.

As the children reached adulthood, they were versed in the art of ambush, murder, and cannibalistic butchery. For years they continued, reaching adulthood, growing older, more savage, and less human in most respects. Twenty-five years passed, with scarcely a clue—other than an occasional piece of pickled human meat which has spoiled, being found on the sea-strand or elsewhere—as to what was going on. Cannibalism was at last suspected, and searches of the countryside were made. Many were wrongly accused of the crimes and even executed.

All in all, it has been estimated that over a thousand people died at the hands and gullets of Bean and his filthy family. At the end, their descent into depravity—or the emergence of some horrible and unhuman characteristics—was displayed.

One night a band of Beans waylaid a man and woman who were passing near their domain. The Bean females fell upon the unfortunate woman like animals, cutting her throat and sucking at the bloody wound even as she died. They also disemboweled her on the spot, and in the manner of hyenas, began to feast upon her entrails and organs before the eyes of her embattled and horrified husband.

The sudden appearance of a band of twenty or more riders frightened the Beans away, and the husband escaped their clutches. They had attempted to drag his wife's body toward their lair, but had abandoned it some distance away, and the man showed this to the riders and told his tale. He was taken to Glasgow where he made his report to the authorities, and after a few days the King himself arrived on the scene, accompanied by four hundred soldiers who were each eager to dispatch the human monstrosities described.

At first the company passed by the low-entranced cave, searching for tracks along the strand; but the bloodhounds they'd brought along had a different idea, and set to howling and barking at the cavern entrance. The soldiers fetched torches and entered the black den. As they descended, they beheld a horrid spectacle that could have come from the most hellish myth: the bodies of men, women, and children, dismembered and hung up to cure; their limbs pickled and in stacks; and a tremendous amount of stolen wealth, a treasure-trove accumulated over the years. Eventually they came into the living quarters of the Beans themselves, who may have been sleeping off a night's degeneracy or ramblings. There was a standoff, but at last the cannibals surrendered and were taken captive.

The king had them marched in bonds to Edinburgh at first, but they had no trial. They were taken to Leith for execution. The men were dismembered while alive, as their females and offspring were made to watch; then the latter were burned to death, by all accounts screaming curses and blasphemies to the very end.

The cave where these human monsters made their home is believed by locals to be haunted to this day...and has never been fully explored.

Here we are back in Scotland; here are the deep caverns going down, down, down into the unknown crust of the earth. Like so many other groups with apparent underworld connections, we find a bit of inbreeding as well. This of course raises questions: Is inbreeding responsible for the creation of such aberrant forms of man? Or does inbreeding create an isolated gene-pool in which certain genetic traits become overly-dominant, traits which may be desired by the unseen forces of the underworld? This story is filled with the hint of unknown and even unguessable mysteries and horrors. What might have come out of the depths of the world on cold Scottish nights, to bargain with Bean in the cavern, or to procreate with his wife, his sons and daughters, grandsons and granddaughters? What hellish traits might have been passed on to them, traits which craved or delighted in the tastes of blood and the "sacrificial" organs? Where the offspring of Sawney Bean completely human beings, as the term is currently defined?

Current theories about the formation of the ultimate sociopath—a serial killer or a cannibalistic serial killer—state that, while environment and childhood trauma may have some sort of triggering effect (only in some cases), that many of these perpetrators are simply "wired incorrectly." Their brains, their emotions, and their cold cunning are not human, are not even mammalian. They appear more or less normal on the outside, but inwardly they are essentially the same, in terms of compassion and other human emotions, not to mention predatory instincts, as a komodo dragon.

Goya captured the horrors of cannibalism in his painting depicting Saturn, devouring his offspring.

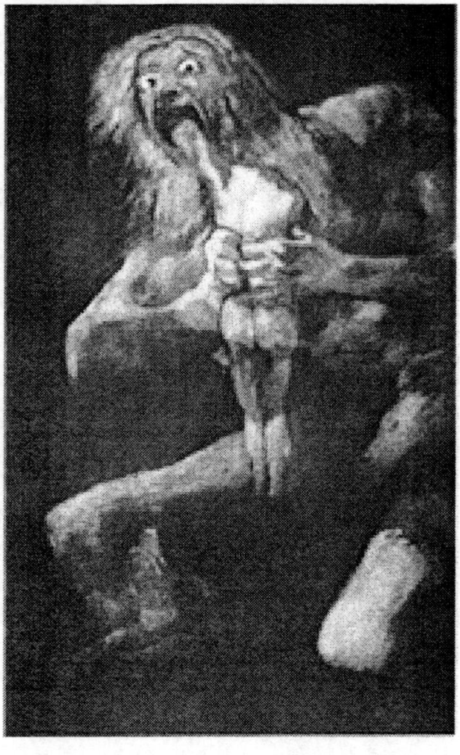

(See Color Inserts)

According to the leading expert of serial killers (and the coiner of the phrase), FBI Investigator Robert Ressler, childhood environment and abuse and resultant sexual dysfunction play a part in the development of human monsters, and this may be true in many instances. Yet he also states,

> *Maybe bad genetics is somehow triggered by a bad family environment. I don't know. Nobody knows. We haven't been able to deal with pure evil scientifically.*

Many scientists, however, believe that the genetic connection—whether a malfunction or other inexplicable difference—is the most likely explanation for such aberrant and violent individuals. The question remains: Where do such genetic predispositions or anomalies ultimately originate?

Could it be that, due to a contaminated heredity, the worst among us are not really completely of us? Do monsters that are not completely human, in spite of outward appearances, walk abroad in the surface world? This isn't a view that would be readily accepted by modern science, but the circumstantial evidence would indicate that this might be the case. One fact that indicates a strong possibility of cross-specie contamination, is that, as a species, only human beings have over four thousand known genetic defects, while our nearest chromosomal relatives (according to evolutionary science), the great apes, have almost none.

Various mythical and folk traditions around the world make it quite clear that demons, ogres, monsters, and so forth have always been with us in human guise. Did the various ancient traditions know more about the truth than modern humanity, science, and law-enforcement would care to admit? Entertaining such a radically non-politically-correct view is to invite scorn and ridicule, but requires both courage, and the serious contemplation of facts and circumstantial evidence which history and folklore provide us.

History is filled with their names and their shadows: Sawney Bean, Jeffrey Dahmer, John Wayne Gacy, Andrei Chikatilo, Ted Bundy, Gilles de Raise, Pedro Alonso Lopez (the "Monster of the Andes"), Richard Trenton Chase, Gilles Ganier, Peter Stubbe, Elizabeth Bathory, Ed Kemper. And patterns emerge. They victimize and rape children, or young men and women. They often eat human flesh and organs, and drink or bathe in the blood of their victims. Some engage in necrophilia as well as necrophagia. Many seek out locales or environments which are reminiscent of the subterranean realm, in order to hide, store, or dispose of the remains of their victims. Bloodlust is particularly overpowering for these humanoid animals, and like the underworld monsters of old, they are fascinated by the organs that would have once been reserved for sacrificial rites. Richard Chase, the "Vampire Killler of Sacramento," pureed the blood and organs of his victims in a blender, and he and Dahmer both drank copiously of the blood of those they had killed. In an interview at a University seminar, Robert Ressler had this to say about Chase:

> *His eyes really got to me. I'll never forget them. No pupils, just black spots. These were evil eyes that stayed with me long after the interview.*

Which of course would be the light-hungry, dilated eyes of a subterranean dweller in darkness.

So much for theories of environment over heredity. At least to this writer, this sounds like a physiological rather than an environmental result....and now, awaiting trial in Great Falls, Montana, is a newly found-out monster who fits the profile, and to some extent the physical description, of a "dero" almost precisely....Bloated, slovenly, a sexual pervert and torturer, a murderer, butcher, and eater of young boys....What perverse, twisted, and non-human sort of "man" is this? An epicure after the likes of the Sawney Bean clan, this cannibal ogre out of the darkest fairytales, this beast disguised as a man, had even prepared a cookbook in code for his own amusement, listing such dishes as "Sex A La Carte," "Barbequed Kid," "Little Boy Stew," "My Little Kid Dessert," "Little Boy Pot Pies," and "Lunch is Served on the Patio with Roasted Child." It is said that authorities may never know how many missing children went into this degenerate sub-human's home butcher shop and larder.

An ending such as that dealt to Sawney Bean and his spawn would be absolutely fitting for a fiend such as this.

Zoologist Christopher Hitchens may have said it best, when he commented on the interaction between reptiles and mammals in his recent book, *No One Left to Lie To*:

> *The reptile can break into the mammal's nest, destroy and eat all of the young and be burrowed into the still warm and living flank of the mother before any reaction is evident. Our anthropomorphic verdict would be that reptiles don't even know they are lucky, while mammals don't really believe that reptiles can exist.*

Where do our legends of cannibalistic or man-eating giants, trolls, and monsters really come from? And why does this twisted obsession enter the hearts of some 'men?'

Illustration: H.J. Ford, for Lang's RED FAIRY BOOK, 1890

5. WHO, WHAT, AND FROM WHERE?

PROSPERO:
Refusing her grand hest, she did confine thee
By help of her most potent ministers,
And in her most immitigable rage,
Into a cloven pine; within which rift
Imprisoned thou didst painfully remain
A dozen years, within which space she died
And left thee there, where thou didst vent thy groans
As fast as mill-wheels strike. Then was this island—
Save for the son that she did not litter here,
A freckled whelp, hag-born—not honored with a human shape.
ARIEL:
Yes, Caliban her son.

—William Shakespeare, The Tempest

Many strange genetic defects lurk within the genetic code of human beings, yet by no means are those who suffer with them non-human, or monsters of any sort. Only upon occasion does a person of such strangeness and violent disposition shock us all into realizing that there are others, who must surely not be completely of our kind, and as has been seen, these individuals often are indistinguishable, by the regularity of their appearance, from anyone else. Not all have the appearance of a Rene Randolier.

Some of the other defects, unrelated to violent acts or behavior, which may speak of a possible occurrence in the distant past of a subterranean or other-than-human intrusion into our collective gene-pool, are the defects of human albinism, along with another condition called Xeroderma Pigmentosum, or XP. The latter condition is a rare genetic defect which leaves lacking the sufferer's DNA repair mechanisms for repairing the cellular damage caused by ultraviolet radiation, particularly sunlight. DNA damage is always irreversible. Fewer than one thousand XP cases are documented worldwide. The condition is life-threatening, and can result in the following symptoms: blistering, freckling, and second- and third-degree burns from brief exposure to sunlight, lack of cellular repair or regeneration after burning has occurred, blindness and deafness, dwarfism and hypergonadism, increased skin and eye cancers, mental retardation and other conditions such as Ataxia-Telangiectasia, Bloom Syndrome, Cockayne's Syndrome, Fanconi's Anemia, and Trichothiodystrophy The unfortunate sufferers of XP tend to be the most innocent of our kind, children, and life for these poor kids is seldom long or pleasant. They have to remain in semi- or complete darkness, shielded from the light of day at all times, and are allowed to venture out without protective gear only at night. As long as they're protected from UV radiation, these children are, in many cases, the same as any others. Has their frail humanity been contaminated or compromised by a genetic trait from another hominid or humanoid species? Fairies, trolls, vampires, ghuls, evil beings of all sorts, the deros and teros of Richard Shaver: All of these are said to shun the light of the sun for fear of the damage, injury, or death it will surely bring upon them. Does XP represent genuine medical proof that there may be more to these so-called myths and fantasies than we realize?

Some traits may have entered the human gene-pool more recently. The carnival performer Grady Stiles Jr., the son of "Lobster Man" Grady Stiles, Sr., and known by his stage-name of "Lobster Boy," represented the fifth generation of his family to suffer from a strange condition in which his hands and forearms were blunt, two-pronged, lobster-like claws or clubs. Instead of legs, he had seal-like appendages, reminiscent of aquatic flippers. This condition was not simply a "birth-defect," but represented a set of inherited traits, as it had been passed on successfully for five generations, and finally to a sixth. Where on, above, or under the earth or sea did these traits come from?

While well-known or famed on the "carny" circuit, Grady led a personal life that was marked by brutality, cruelty, rage, and murder. He married, and raised four children, two of whom inherited his physical traits to some degree. When one of his daughters sought to escape his physically and emotionally abusive control by eloping in 1978, he shot his would-be son-in-law at point-blank range, murdering the young man in cold blood. Yet at his trial he and his defense played so well upon sympathies for his condition, that he was acknowledged as having committed the murder, but was never jailed! After this he divorced, remarried, and moved to Gibsontown, Florida, a small community of carnival performers and so-called "freaks." Here he set out to spend the remainder of his life pursuing one of his favorite pastimes, physically abusing his family. He was quoted by them as often have remarked, "I killed once and got away with it. I can do it again."

As a result of the extreme beatings he inflicted upon his stepson and wife, they hired a neighbor to murder him in 1992 and he was put out his own misery, and out of theirs as well.

"Surely," skeptics will remark, "his condition was due to spontaneous genetic mutation," resulting from "the exposure of his ancestor to a radioactive substance or other source of radiation." This explanation, however, does not hold water, since evolution holds that spontaneous mutation occurs under pressure and to serve a function, or to increase genetic and procreative viability and survival. Nor does it explain the strength of the genetic traits, as they were passed down for six generations. In terms of the stunted yet powerful limbs, there is some resemblance to descriptions of "the Monster of Glamis." Grady's form or condition was decidely aquatic in terms of its non-humanoid features. So, other than in the myths and legends of lusty mermaids and mermen, elemental undines, and the randy Quinotaur rapist of Merovingian legend, where do we find more recent clues as to the possible origin of the offending and invasive genetic code?

On October 12, 1973, an event which took place in Pascagoula, Mississippi may have revealed some clues, after a fashion.

While fishing off an old pier near the Gulf of Mexico, on the Pascagoula River, Charles Hickson and Calvin Parker, aged 42 and 19 respectively, had a close encounter of the most unpleasant kind. An egg-shaped craft approached them, immobilized the pair somehow, and took them aboard the craft. These humanoid creatures were described as being about five feet in height, covered in wrinkled, gray, "elephant-like" skin (a suit to protect from the effects of solar radiation?), and had a large, long

LOBSTER-CLAW APPENDAGE at the end of each arm. Their legs were described as "fused," more like one appendage than two, and they didn't use this for locomotion, as they floated or levitated everywhere they went.

Grady Stiles, Jr., aka "Lobster Boy"

A medieval likeness of a "Seducer from the Deep."

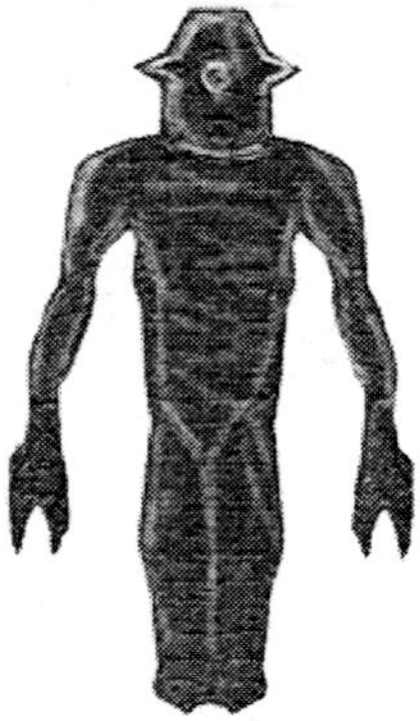

Image of description of a "Pascagoula" Ufonaut.

Details are sketchy as to what took place onboard the UFO, but as-yet-unrealized "missing time" might account for that. Be that as it may, the men were eventually released (after the "aliens" obtained that for which they had come?), reported the encounter and abduction to the proper authorities, and both passed lie-detector tests. One of these men has suffered at least two nervous breakdowns as a result of this encounter, and it is considered one of the most reliable and substantive UFO encounters, or CE4s (Close Encounter of the Fourth Kind) to date.

It is of course assumed by some UFO researchers that the object and its occupants originated "somewhere else," i.e., another planet or astronomical location. This is not a logical assumption, given the proximity of the Gulf of Mexico (a hotbed of UFO sightings) to the scene, and the long history, worldwide, of the sightings of and encounters with USOs (Unidentified Submarine Objects) which turn into UFOs upon leaving the water (as discussed at some length by this writer in "The Deception of the Depths: Humanity and the Ocean of Secrecy"). Here we have humanoid beings, possibly on a mission to obtain genetic material or samples, and these beings have lobster-like claws or arms, and a seal-like or "fused" lower extremity (much like the lower part of an Egyptian sarcophagus in form), possibly to immobilize "dangling" appendages. Other so-called "extraterrestrials" or aliens have matched this description, to greater or lesser degree, as well. Among these are the pincher-packing "Metaluna Monster," (Argentina, 1957) and dozens of others which have been described as having hands which were "mitten-like." This in turn recalls the centuries-old tradition of what Katherine Briggs called "the deformities of the fairies," which were often similar descriptions of fairy, troll, gnome, etc. hands and feet.

It would be interesting to ascertain, if possible, whether the ancestors of Grady Stiles, Jr. ever had such an encounter as well, with suited beings in broad daylight, or uncovered ones in the dead of night...but this will probably never be known.

AFTERWORD

"Come listen, my men, while I tell you again
The five unmistakable marks
By which you may know, wheresoever you go,
The warranted genuine Snarks."

—Lewis Carroll, "The Hunting of the Snark"

"We must not look at goblin men,
We must not buy their fruits:
Who knows upon what soil they fed
Their hungry thirsty roots."
"Come buy," call the goblins
Hobbling down the glen.
"O!" cried Lizzie, "Laura, Laura,
You should not peep at goblin men."

—Christina Rossetti, "Goblin Market"

The suppositions postulated or ruminated upon in this examination will infuriate and offend some; others will scoff or shake their heads in dismissal. The premise is not designed to offend, nor to bring scorn or mistreatment to those unfortunates who suffer from conditions beyond their control. This is simply an examination of facts and patterns, patterns which become very clear if looked at in an objective fashion. If the human gene-pool, as evidenced by the amount of errors or illnesses in our collective genetic code, has been invaded by chromosonal data from a non-human (non-Homo Sapien, non-Adamic, or non-Cro-Magnon) source, then we may ALL have been compromised to some degree, long ago. Or perhaps not, and our ever-increasing population on this planet explains in part the sudden upsurge in UFO and paranormal activity. A feast, a cornucopia of human genetic variation is now available, and the unseen parasites, dependent upon it as they have been for ages, are rushing to the beat of an agenda of their own.

We must not "buy their fruit," or the tons of propaganda which permeates our waking, sunlit, media-driven world, to mollify and nullify our fears and suspicions. Such fears are primal and ancient, with good reason to be in place—they are instinctive, and have insured our survival as a species for a very long time. The various yet interrelated opportunistic intruders are not benevolent, or "guiding our evolution," nor are they "time-travelers" who are our own descendants, as many have dreamed, or swallowed whole. They are not from distant suns or galactic realms, but are native to this solar system and most often, to this planet (worlds within worlds), as many time-honored traditions attest. Disinformation is a powerful tool, and a heavy propaganda campaign is underway. It is perhaps time to look to elder legends and traditions, and to proven, protective religions and belief-systems, for information and for spiritual strength; if these are balanced with an aware and serious scientific point of view, and a lack of fear in the face of diabolic opposition, humanity will prevail. We may have sometimes forgotten the rules of the game, but somehow we manage to win just the same.

Let us be bravely aware of our victories as they occur.

Part V: EARTH, MOTHER of MYSTERIES

(See Color Inserts)

1. A HIDDEN BIOSPHERE

What's going on down there–and all around us? In January 2001, a group of spelunkers explored a cave called Cueva del Arroyo in central Mexico. After descending a total of four times, the cavers found themselves deep below the ground, in a chamber with several streams crossing its muddy floor. Distinctly visible–and photographed–were a series of bipedal footprints in the mud, leading first down to the water, and then back away and into the depths of the Earth. It was estimated, from the size and depth of the tracks, that the unknown creature weighed about 18 pounds and walked upright, moving in total darkness (More information can be found at: http//www.telegen.com/timefor.html)

Those photos raise many questions: Where did such a thing come from, and to where did it return? What is it? What does it eat? Is there a breeding population? Is it wandering up from greater depths? Obviously, it wouldn't come in from outside the cave to feed, particularly to a location so deep. (It should also be noted that the arroyos outside of the cave system are arid and almost devoid of life, and food.)

Another recent incident took place in Pitrufquen, Argentina. Scott Corrales recounts the story in an article in the Diario Austral de Temuco article on February 18, 2001, under the heading "Local residents terrified of 'Imp.'"

> *According to witnesses, a small entity only twenty inches tall, but possessed of both unusual strength and malice, has been terrorizing children and adults alike, exsanguinating (draining the blood from) chickens and livestock, and generally creating a panic. The creature, described as having " unusual " ears, making a noise like a baby crying, and wearing a blanket-type article of clothing, is believed by the local citizens to be a diablito, or little devil (imp). He is said to emerge from or live in a local cave, and is described as " a little man with wizened features, hairy pig-like ears and shining eyes." Children have been throwing stones at this creature in spite of their fears; neighborhood dogs won't go near it.*

Like many other unknown entities (the Hopkinsville Goblins, El Chupacabras, etc.), "El Diablito" is described as being both impervious to attack and unusually strong. If this is a being who either evolved in or was designed for an environment which is under much greater pressure than our surface world, this difference in density and physical power would be explained.

2. MYTH AND LEGEND

Evidence for alternate biospheres (or unknown aspects of the one in which we live) exists both in eyewitness accounts of certain unknown creatures and within the realm of mythology.

The legend of Cadmus recounts the story of a hero who was not only the inventor of the alphabet and founder of the city of Thebes, but was also a warrior of renown. He battled a terrible dragon that came out of the earth, killed it, and then buried its teeth in the soil as he had been instructed. From the planted teeth grew a host of armed warriors, who fought one another until all but five were dead.

These men served Cadmus and help build the citadel of Thebes. From these five "dragon-born," and human women, were descended the five noblest families of Thebes–families who ultimately were descended from the dragon and the earth. The symbolism is obvious: the nobility of Thebes were descended from the dragon-race or nation, representing a link between the human and nonhuman worlds.

3. DRAGON SPOUSES

Ancient Chinese mythology gives us the tale of the dragon-husband, who extorted a wife from a farmer with seven daughters. Not wishing to be eaten, the farmer acquiesced, and the dragon took the youngest daughter to his "kingdom beneath the sea," a cavern-realm. Here the girl witnesses the dragon-king dancing with delight, "curling" like a ribbon (or a DNA helix, a reproductive symbol?), and transforming into a "handsome prince" or king.

The human wife eventually longs to visit her family, where a jealous sister tries to drown or otherwise subdue her and take her place as the dragon-wife.

Some versions of the tale begin with a giant snake in the dragon role, a snake which eventually transforms into the dragon, once the marriage is made, and then into the humanoid form. A parallel to this story is found in several Japanese legends, where the human participant is generally male, and the dragon-wife is the daughter of "the dragon-king of the sea."

Japan is also the location of Lake Tazawa, which never freezes in winter in spite of its frozen Akita location. Legend maintains that this non-freezing quality is due to the vigorous love-making, at the bottom of the lake, of the great beauty Tatsuko, and her abductor, a dragon-husband of great wealth and power.

A similar legend surrounds Lake Biwa, north of Kyoto. Biwa is said to be the home of a dragon-king with many wives, but only one daughter, who is called Benzai-Ten. Benzai-Ten fell in love with a mortal man, and lavished many gifts upon him. Benzai-Ten is described as beautiful (yet able to change into dragon-form), with eyes of deepest, shiniest black.

In these legends of the Orient, and many others from the same region, the nobility of the underworld realm lust for, or fall in love with, ordinary human beings of humble stock and human nobility alike. In Benzai-Ten we get a picture of the dilated black eyes of a dweller in a lowly-lit environment.

4. WATER-CATTLE

Men and Gods in Mongolia, by Henning Haslund, originally pubished in 1935 (new edition 1992, Adventures Unlimited Press), repeats a Mongolian's account regarding the lake called Socho-nor in Outer Mongolia (Chapter XV, pgs. 137-138):

> *Where the lake of Socho-nor now heaves there once stood a great and wealthy city. Its inhabitants lived a sinful and unrighteous life which called forth the wrath of the gods. In order to punish the people the angry gods one night caused the whole city to sink into the earth, and in the hollow made by this they created a lake into which they cast alkali*

> *and salt so that no people would be able to live in its proximity. But the innocent cattle were not to suffer for the sake of their sinful owners and the animals therefore were transformed into socho, the water-cattle, who could live and multiply on the rich pastures of the lakes bed. Sometimes one could see them rise to the surface of the water, but always right out in the middle of the lake. Sometimes, when the water level was especially low, early morning hunters had seen their tracks on the brink, and old folk could remember their parents telling them how sixty years ago, when the lake was dried up after a very dry summer, the mysterious water cattle had been seen disappearing into the interior of the earth through a hole in the bottom of the lake.*

This parallels legends about “water-cattle” and “water-horses” in the folklore of the Scottish Highlands and Ireland, as well as in the Norse world. In these traditions, “lake and sea/fjord monsters”–the same entities or creatures chased currently by modern cryptozoologists, and after which the vikings designed their dragon-prowed ships–were said to dwell in cavern or “faerie” lands beneath the bottoms of lakes and seas. Another parallel is found in the Mongolian version, which, like the many Celtic versions, states that interbreeding takes place upon occasion between the lake-creatures and terrestrial cattle, resulting in a form of “hybrid vigor:”

> *And if this poor huntsman on the next night led a cow to the edge of the lake and laid himself down to sleep beside it after having cried the proper formula out over the lake, during the night the cow would be visited by socho and soon afterwards would become pregnant and give birth to a calf whose productivity was equal to that of a herd of many hundred animals.*

These motifs seem to be universal, regardless of race, culture, or location. This tale also puts one in mind of recent discoveries in Loch Ness, for not only are the sides of the loch pocked with unplumbed caverns, but there has been a recent discovery of a huge “hole” in the bottom of the loch, which leads away into some subterranean region. This hole is located at a depth of 700 or more feet, and is nearly impossible to explore at present. Could this be a major source of ingress and egress for the unknown animals in the loch, through which they access the sea, or even unknown subterranean realms? Given the nature of the Loch Ness "mystery," this is a logical hypothesis.

In recent years many other anomalous events have occurred which would indicate the existence of an underworld biosphere of some sort. In addition to the multiple associations of hairy humanoid or hominid and various "chupacabras" forms (ranging from humanoid to dog-like) with caves, mines, tunnels, sewer-systems, etc., other forms have been noted as well.

(See Color Inserts)

5. THE MOHOROVICIC DISCONTINUITY

Evidence mounts which indicates the existence of a vast underworld biosphere or ecosystem beneath our feet, beyond the topmost caves and caverns. At great depths there exists an anomalous region which is called the Mohorovicic discontinuity, or Moho. Named for Yugoslav geophysicist Andrija Mohorovicic (1857-1936), who deduced its presence after studying seismic waves from the 1909 Kulpa Valley earthquake, the Moho follows the variations in the Earth's crustal thickness, and lies at a depth of approximately 32 km or 20 miles beneath the continents, and about 10 km or 6 miles under the floors of the oceans. Marking the boundary between the planetary crust and mantle, this region is marked by seismic irregularities and variations in the speed of earthquake waves, which indicate either a hollow region with peaks (mountains) and troughs (valleys), or a region which is filled with some sort of material which is less dense than stone or magma, like water. Essentially, it is a cavern-world, perhaps with both water and air regions, and one which is currently out of the reach of human exploration. But there are even deeper, wetter regions which might support or engender life.

According to the October, 1994 issue of Discover Magazine, Princeton seismologists G. Nolet and A. Zielhuis have postulated, based on extensive data interpretation, that a gigantic reservoir of water exists about 900 kilometers under modern-day Europe. Apparently this "ocean" or "sea" was subducted in some way, for about 400-500 million years ago a surface-world ocean existed in the same region (Zimmer, Carl; "The Ocean Within," Discover, 15:20, October 1994).

New Scientist magazine, September 1994, states that Washington University's H. Wysession has discovered a water-rich slab or stratum, approximately 250 kilometers thick, existing at a depth of 2,700 kilometers beneath present-day Indonesia. Wysession believes that this huge body of water may be the subducted floor of the paleo-ocean called Tethys, vanished from the surface world for 250 million years (Redfern, Martin; "Lost Ocean Found Deep in the Earth," New Scientist, p. 16, September 3, 1994.).

The ability of various "cryptid" and seemingly paranormal creatures to pop into and out of our reality might in part be explained by the use of concealed or even unnoticed entrances which could lead down to these unexplored regions. The habit of many UFOs which plunge into the oceans of the Earth might also make more sense, as the Moho is much nearer the floors of the oceans and might be more readily accessible; and the aforementioned "lake monsters" could take a more circuitous route by means of subterranean waterways, aquifers, and so forth (as expounded on at length in "The Deep Dwellers", the first section of *Caverns, Cauldrons, and Concealed Creatures*).

6. THE ANCIENT ENEMY

Western and Middle Eastern traditions have long held that there is an unknown aspect to our biosphere, in which an ancient enemy lies imprisoned. From this prison, these eternally-bound beings send forth their servants, demonic, partially-human (or genetically engineered; take your pick), to do their mischief in the surface world. In these various traditions of horned devils, horny djinn, and deros and teros, we see echoes of even more ancient tales of an earlier age.

Many researchers have disclosed cases of "Horned Giants" which have been found all over North and Central America, along with midget or dwarf-like mummies, doll-sized mummies or bones, and other strange humanoid remains with double rows of teeth and extra fingers and toes. These things have been verified, in print at the time, as having been found, but all seem to "disappear." David Childress provides some excellent accounts of this pattern of chicanery and "doing away with" relics, as well as who the possible culprits of this archaeological cover-up might be, particularly in his book "Lost Cities of North and Central America."

According to esoteric traditions, the "horned giants" were the direct offspring (by sexual procreation rather than genetic engineering), with human women, of the nephilim–or titans, or fallen angels–take your pick. The Apocryphal *Book of Enoch* indicates that, acting on behalf of their nephilim masters/fathers, these giants ruled the antediluvian world with terror and cannibalistic savagery, preying on humanity. The archetypal "demons" and "devils" of humanity have traditionally sported horns, and the horned god Cernunnos had a taste for an occasional barbeque flesh. Apparently the race of giants were hybrids of a sort, but not genetically engineered in the same sense as the manticore/chupacabras, or Sumerian PAZUZU or UTUKKU (Mothman) and other types.

The Hebrew word nephilim or nefilim did not really mean "giants," as this was a mistranslation into English by King James' boys. Nephilim means "those who descended," "those who fell," or "those who came down," meaning of course, from the sky. The Books of Enoch describe them as a second band of "fallen angels," 300 in number, who "looked on the daughters of men and found them fair–" and then some. These were either spiritual, cosmic, or extraterrestrial beings, and they were the second wave of fallen ones to come to the Earth, in this case simply as a result of their lust for human women (not so different from current "UFO abduction and hybridization" accounts). Their original function was as "Watchers," and they were supposed to watch over the Earth and its inhabitants. From this perspective, their transgression was akin to rape and child sexual abuse.

The nephilim apparently broke some serious rules when they did this, and the result was the race of "the mighty men of renown," from the Torah, described in the Apocrypha as cannibalistic, sadistic giants. Hybrid vigor, to say the least. Other crimes before "heaven" that the nephilim practiced included taking the daughters of men to be mothers of hybrid children.

The *Ethiopic Book of Enoch* states:

> *"And the angels, the children of heaven, said unto themselves, let us choose wives for ourselves from among the daughters of men, and they will bear us children."*

Note that both the nephilim and their children are generally considered to be Nephilim, just as the Anakim (sons of Anak) bore the name of the Anunna or Anunnaki, who were their forebears.

The nephilim also conducted genetic experiments and gene-splicing using all creatures, including human beings. They "mixed the man with the horse, and vice versa," and "transferred one embryo from one womb to another." The result of all of this was a myriad of monstrosities which preyed upon humanity: "For they sinned against all flesh, all manner of fish, and birds, wild animals (mammals) and reptiles," creating chaos and terror on the Earth. Times got tough on Planet Earth. The Enochian Books state: "The giants (children of the nephilim) consumed all food that men had, so that people despised them and refused to feed them. So the giants preyed upon the people, to eat them."

7. ENOCH

Enoch was a human man, chosen by the heavenly court to be the prosecutor and judge in the trial of the nephilim, where they were called to account for their crimes, for they had "defiled themselves with women, and like them, produced creatures of blood and flesh that would someday die." Apparently this was a big no-no in the eyes of the heavenly (or galactic) Father.

Enoch was supposedly chosen for his familiarity with the case and facts at hand, being a human being from the planet in question, and also for his intelligence and learning, which the non-fallen Watchers themselves had taught him throughout his life. The Book of Enoch reports his spoken verdict:

" Now you receive Judgment. From now on you will not be able to pass beyond the heavens and into eternity, but will remain INSIDE THE EARTH, imprisoned for all time. Before that [the execution of the sentence] you will watch the destruction of your beloved offspring." Meaning, of course, the hybrid giants, and the various genetically-engineered forms as well:

" Destroy the children of their adultery and exile the Watchers from among the heavenly host, and send their children against one another in war, so that they might fight and be killed."

This then was followed by what is traditionally known as the Deluge, which wiped out most of the genetic taint and aberrations on the Earth. Note that Noah and his wife and children, and his children's families, were chosen to be saved or spared "because he was without blemish." This is not just a reference to his obedient character, but is also a reference to the fact that his tribe or family represented one of, or perhaps the only, group of human beings who were untainted by decades or centuries of interbreeding with the nephilim and their "children." Both of the latter are the Titans of Greek myth as well, unable to ascend again to Olympus (the heavenly or galactic kingdom, whichever you prefer), or beyond a designated point "up there" somewhere. Traditionally, the Hebrew and Sumerian cosmologies provided for a number of heavens, nested like concentric spheres, with the ultimate expanse " beyond" the stars and such being the realm of God or the gods.

The Pseudopigraphal *Book of Jubilees,* Chapter 5, tells us:

> *6. And against the angels whom He had sent upon the earth, He was exceedingly wroth, and He gave commandment to root them out of all their dominion, and He bade us to bind them in the depths of the earth, and behold they are bound in the midst of them, and are [kept] separate.*

and

> *10. And their fathers were witnesses [of their destruction], and after this they were bound in the depths of the earth for ever, until the day of the great condemnation, when judgment is executed on all those who have corrupted their ways and their works before the Lord.*

As for bigfoot, sasquatch, yeti, and related forms (the Hebrew sa'ir, or "hairy one," was said to haunt the deserts and "wild places,") there's little doubt that these are pre-human forms which probably have early relatives well displayed in the fossil record (gigantopithecus especially seems to fit the description, but others might as well). These beings tie in to everything else, and are often represented in some traditions as mute and animal-like servants of the nephilim, anunnuaki, dragon-kings, and other beings who were "banished to the underworld," " the pit," Tartarus, and so on. There is ample evidence to support this contention, ranging from European to Hindu legends, from the Middle East to the tales of Native Americans.

8. ALIENS OR HYBRIDS?

As far as the wide range of seemingly "alien" forms that abuductees and experiencers report, we should stand back and view these with a critical eye. It should be noted that there is always a great deal of effort put forth to convince abductees, contactees, experiencers, witnesses, and so forth that the UFO entities of various types "come from somewhere else."

"We come from Zeta Reticuli," they claim—or Orion, Epsilon Bootes, another star system, or whatever. But the overwhelming evidence would indicate that the UFO phenomenon is localized to the region of the Earth, and this has been the case for thousands of years. Not just to the region of the Earth, but of the Earth. With the possible addition of our moon and Mars, the Earth seems to be the primary source of encounters with, and object of interest of, these beings. Earth is "where the action is," where the genetic wealth is, where the various battles between different schools of thought, methods of interaction, religions, philosophies, and quite possibly the "big mystery" (spiritual truth or ultimate destiny) is to play out or to be revealed. In short, evidence simply indicates that this is the case. UFO entities are no strangers to manipulating human religious tendencies or belief-systems (all systems) and have done so repeatedly.

To quote from my own work briefly ("The Deep Dwellers" , from the book *Caverns, Cauldrons, and Concealed Creatures*):

> *Yet the biology of all the different types or castes of abductors, as horrific as it might appear to superstitious human eyes, is essentially that of animal forms which are natural to the Earth: mammalian and reptilian. Obviously, it is very important to these beings that such a logical connection not be made. If there is even a shred of truth to UFO abduction accounts, then it is more than apparent that the abductors want their victims, and humanity at large, to believe that they are from "somewhere else." While humanity looks continually upward at enigmas in the sky, what is transpiring beneath our very feet?"*

It should indeed be noted that an earthly vertebrate template is almost always in effect with such "alien" creatures: they tend to be bipedal, with two arms, two legs, two eyes, one head, one central body or trunk, binocular (predatory) vision or ocular placement, et cetera. They are only occasionally insectoid, but these are probably vertebrates as well, since two arms and legs only are noted in these cases.

Logic and numbers would indicate that this pattern would simply not repeat over and over throughout a universe as vast as ours. The range of forms would have to be more numerous. Despite the strangeness or outlandish nature of their initial appearances, UFO entities and abductors ("humanoids") are not so dissimilar from human beings, with earthly animal characteristics often described.

After many years of researching these interrelated phenomena, it has become apparent to this writer that the modern abduction phenomenon is the same as the ancient, medieval, and all other "abduction phenomena" which have taken place over the centuries, and which seem to blur the lines between physical reality and various types of "altered states" of awareness. Shamanism's spirit abductors, alchemy's elementals, Christianity's demons, the fairies of rural folk traditions, the various abducting and seducing demons of Islam, Hinduism, Buddhism, and pretty much all other religious or mythical traditions exhibit the same types of behavior, preferences, similar appearances, and so on.

In other words, the UFO abduction scenario is just a new mask for an old player, a new get-up for forces which have always plagued or hornswoggled us for their own selfish purposes. A perfect example is the meaningless "star-map," drawn on paper and hanging from a wall, which was shown to Betty and Barney Hill. Why would such technologically advanced beings be using a drawn or printed map, containing no recognizable star systems or constellations, and make a point of showing it to their captives, stating, in effect, "We come from here?" Where did they get the paper? Why not pull up a hologram or a screen?

9. COSMIC DISINFORMATION

This is identical behavior to the lack of use of anesthetic and so forth in their so-called medical experiments, which actually blend sadism and genetic theft activity. The Hill "map" scene, carefully contrived, was pure misinformation, given to captives who were in some sort of altered and highly-suggestible state of mind, so that they would retain the notion that their abductors "came from a far-away star system, 'cause they showed us a map." Meaningless in terms of the information it supposedly conveyed, but not meaningless in terms of the patterns of behavior exhibited which are indicative of an ongoing deception and manipulation.

We are dealing with a phenomenon which is very sophisticated, and which at times blurs the edges between what is considered to be reality, and the so-called supernatural. But this phenomenon is very real.

Current advances in genetic engineering, talk of super-smart " biochips " which can be integrated into living brains or nervous systems, stealth technologies of various types- the group or groups under investigation here. They have been around in one form or another much longer than we have. They have a vast array of accumulated knowledge or technology, which for all practical purposes might as well be called “magic.” Maybe it is magic, insofar as their understanding of what makes up reality might so vastly surpass ours that their science can affect the fundamental and underlying (quantum) laws which govern the universe as we know it. This would explain much. As Arthur C. Clarke’s “third law” states so succinctly:

“Any sufficiently advanced technology is indistinguishable from magic.”

This in no way negates or denies the more logical possibility that “they” are from here and always have been (or at least a very long time indeed), and may have themselves branched out to the immediate neighborhood of close planetary bodies.

Instead of looking off-planet for the sources of many of the mysteries and terrors which have haunted us through the ages, perhaps it’s time for humanity to take a logical look at an origin much closer to home ... beneath our very feet.

Part VI: ENTITIES, ELECTROMAGNETISM, and EMPs

A Study of Correlation Between Types and Characteristics of Anomalous Beings

The Western view of the underworld, or Hell, has much in common with traditions from around the world: the abode of maleficent beings, demons, and monsters, and the place of torment for the human unrighteous. In western religious thought, this view has changed little from Medieval and Renaissance times, until the present.

According to Western tradition, the fallen angels are imprisoned within the Earth, along with the nefilim, a later band of transgressors who lusted after human women. This theme is paralleled from the imprisonment of the Titans and Jotuns, to the modern confabulations of Richard Shaver.

Scenes from The Damned and the Elect, and The Elect Being Called to Paradise and The Damned Being Plunged into Hell (1499-1502), by Luca Signorelli

and

Fall of the Rebel Angels (1620), top center, by by Pieter Pauwel Ruben

(See Color Inserts)

1. AMBASSADORS FROM HELL

The 1542 manuscript, Cabeza de Vaca's *Adventures in the Unknown Interior of America*, contains an account which may sound more than a little familiar to the modern fortean or UFO researcher.

In the 1530s and 40s, the Spaniard Cabeza de Vaca and his conquistadores, following the Mississippi River as they explored the North American interior, came into contact with many Native American tribes of varying dispositions, ranging from hostile to friendly. Upon meeting and befriending the Avavares, possibly in the vicinity of the Ozarks or further westward, the Spaniards were told a most unusual tale by their hosts.

According to the Indians, approximately sixteen years earlier the region had been visited by another "bearded" character of unusual aspect. This fellow was called by the Indians "Badthing," or "Malacosa," an accurate appellation given his proclivities. Even though bearded, his facial features were "never seen clearly" or where otherwise indistinct or else clouded and concealed from perception or memory; he invaded Indian homes by night, accompanied by a "bright light," and the hair of the inhabitants would "stand on end." Apparently paralyzed against resisting, the poor tribespeople would stand helpless, as, armed with a "blazing brand" or wand of some sort, the creature would rush inside and perform intrusive surgery upon its victims. This ranged from abdominal intrusions and taking of intestinal samples, to incisions or perhaps partial amputations of arms and legs, which would then be surgically stitched or otherwise repaired. Apparently, as is the case with modern UFO abductions and medical procedures, anesthetic was not a part of the program.

The androgynous ambiguity of this creature was not lost on the Indians, for they said that during this season of visitation, "Mr. Badthing" would appear in the midst of their feasts and celebrations, dressed alternately as an Indian man or woman, which raises the question of a possible quest for genetic material based on seduction. The Indians offered him food, but he never ate; they inquired as to his place of origin, and Badthing pointed to a nearby crevice in the ground and told them that his home was there, in the regions below.

The Spaniards laughed at the account of the barbarian bogey-man, until the Indians brought forward numerous people who still bore the visible scars of the intrusive or amputation and reattachment surgeries.

Shocked, de Vaca and his men informed the Native Americans that the creature, whom the Indians still greatly feared, was diabolical in nature, but as long as the Christian Spaniards were their guests, it would not return, nor be able to inflict injuries as it had done previously. The Indians were greatly overjoyed at this news, and the Spaniards then instructed them in the rudiments of Christian theology before departing.

Who or what was this "Mr. Badthing," or "SeÒor Malacosa?" Like many modern UFOnauts who perform intrusive surgical procedures, he or it was androgynous or, more likely, sexless, its visitations accompanied by a "bright light," and "electrical sensation," a form of paralysis in the victim, the use of a glowing "wand" or baton, and the taking of tissue or genetic samples. Recent human mutilations, particularly a well-documented case in Brazil, found the victim to have been somehow disemboweled completely through a tiny incision near the navel. The similarities to "Mr. Badthing" are more than coincidental, they are nearly identical. Another interesting point is that the location where all of this occurred may be approximately the area near Cushman, Arkansas, near the final home of Richard S. Shaver, and the place where inner-earth researcher Charles Marcoux met his unusual death from yellow-jackets encountered deep with a cave. This happened as he made an attempt to explore "Blowing Cave," an alleged entrance to the underworld realm.

Some recent UFOlogical researchers have pointed out a preponderance of UFO abductions among people of whole or partial Native American descent. Whether this is related at all to the predatory habits of "Mr. Badthing" has yet to be proven, and as his intrusions parallel those of strange underworld and UFO creatures as recorded all over the world, perhaps the case should just serve as an example of how ongoing and intrusive the phenomena has always been.

In our present day, Argentina has been visited by a cave-originating representative of a super-dense, immensely strong hair-covered dwarf or imp-like race, which they newspapers call "El Diablito," or "The Little Devil."

This little creep terrorizes children, pets, and policemen, picks up cars, and grunts, snarls and so forth. He also apparently attacked or approached a policeman and threatened to "take him to hell," upon which the cop apparently left the scene. El Diablito is also credited with besting two full-sized military sentries and tossing one up into a tree.

Of course hairy dwarfs or dwarves are nothing new to UFOlogy or to folklore, but while the Latinos consider him to be an "imp" or "diablito," in terms of appearance and density, strength, and so forth, he more closely fits the Norse description of the dark elves, or the Germanic Nibelungs.

The dero similarity is fairly obvious, as well, as they were often described by Richard Shaver as being "dwarfish" or stunted. The UFO connection ties in with all of these traditions, of course, as the Scandinavian/Teutonic underworld dwarf was a maker of "marvelous" or magical (technological) items, such as contrivances for carrying the gods around (vehicles), weapons like Thor's hammer Mjolnir, Tyre's "mechanical" or artificial hand, and other marvelous treasures.

The "hairy dwarf" UFOnauts do seem to be mostly a Latin American phenomena, although "dwarfish" hairy entities, sometimes called "monsters," have been reported over the years in caves or other underground areas, all over the world. As reported by Timothy Green Beckley in his book The Shaver Mystery and the Inner Earth (Gray Barker Books., 1967., pp. 61-62), and also by the late Charles Marcoux, similar entities were encountered near McCallester, Oklahoma, at the site of an ancient hill or mound, which of course would immediately equate to fairies, elves, trolls and the like in the Old World.

This mound or hill had been the location of many mysterious livestock mutilations, mysterious noises and cries, and the like. The elderly farmer who owned the land considered the hill to be possessed or haunted by evil spirits.

In search of adventure and rumored treasure, three brave fellows decided to check out the hill for themselves. After a bit of searching, they located a small entrance, which they enlarged sufficiently to permit passage. Making their way within, they discovered a large chamber along with a seemingly bottomless shaft, descending into the depths of the Earth. This shaft was surrounded by a spiral staircase of a scale fit for giants.

The three men carefully descended the staircase for hours, until their flashlights began to grow dim. The decided to return to the surface, but as they neared the entrance they were attacked from behind–i.e., from the region of the pit–by a pack of hairy, four and five-foot tall humanoid creatures. These things attacked with claws and teeth, snarling like animals, and managed to latch onto one of the men as the other two escaped. At this point, and as so often happens with discovered entrances to the underworld realm, a cave-in occurred at the entrance, trapping the embattled man inside. Fortunately, he had had the foresight to bring along a .45 caliber handgun, which he used against his assailants. This did not kill them, as they seemed possessed of an unnatural density, but it did slow them down as the bullets apparently "hurt."

Fortunately, his friends quickly dug the entrance clear, and pulled their companion out. To their shock, he was covered with what they described as "yellow blood," a type of ichor from the injured entities inside the hill. After this adventure, the three joined the army to fight in the Korean War, and never returned, as planned, to take up where they had left off.

Researcher Charles Marcoux visited the site many years later, and reported that the entrance was totally closed by "break-down," the rubble of a cave-in.

As recently as April of 1998, in Pisac, Peru, the strangely-mutilated bodies of two people were reported anonymously to the police (Policia Nacional del Peru, or PNP), who found two unidentified dead men with "all of their muscles exposed and not a single gram of fat on them." In other words, Liposuction from Hell.

Local Indians and others no doubt nodded knowingly. Such a predatory humanoid has been known for centuries throughout the Andes, where it is called "pishtaco," or "lik'ichiri." One tribe of Indians, the Aguaruna, believe that these creatures are "half-gringos" or pale in color, live in caves, shun sunlight, and suck the body fat from their victims, as well as sometimes "butchering" people like livestock for their flesh and blood.

Are the "pishtacos" symbolic memories of the Spanish conquerors, who worked the local people literally to death at times? Or are they relatives or representatives of the same race or species of intrusive beings from which sprang "Mr. Badthing?" And do "symbolic memories" murder grown men in our "modern" era, skin them, and drain all of the subcutaneous and other fat from their bodies?

Probably not.

2. HAIRY, HAIRY, QUITE CONTRARY

The ruin known as Hal Saflieni, on Malta, is among the most ancient and mysterious subterranean sites or structures in the world. Apparently neolithic or even older, the "temple-complex," complete with a buried oracle-chamber from which a speaker could project the demanding voice of a god or goddess throughout the entire complex, is officially composed of three levels. Excavators of the site, in the early 1900s, removed the sacrificed and apparently cannibalized remains of over 30,000 people. The remains of these unfortunate souls had piled up over untold centuries.

According to researcher Riley Crabb, Director of Borderland Sciences Research Associates, and as detailed in his printed 1960s lecture "The Reality of the Underground," the remains of these people were of normal human type and stature. This did not explain the many tiny tunnels, carved from solid rock, which led away from the main, "upper" yet still subterranean chambers, and down and away into darkness. These tunnels are only a few feet in height, and would have been utilized by beings of extremely small stature. Yet there are rumors of and evidence for other types of beings lurking beneath Hal Saflieni, and perhaps the entire island of Malta. In addition to the small tunnels, others of larger size, long-since covered or blocked off, once led away into subterranean catacombs and caverns.

According to Richard Walter's article "Wanderers Awheel in Malta," as featured in the August 1940 National Geographic, and as reprinted by Crabb in his lecture, "Years ago one could walk underground from one end of Malta to the other, but all entrances were closed by the government because of a tragedy. On a sight-seeing trip, comparable to a nature study tour in our own (American) schools, a number of elementary school children and their teachers descended into the tunneled maze and did not return. For weeks mothers declared that they heard wailing and screaming from underground, but numerous excavations and searching parties brought no trace of the lost souls. After three weeks they were finally given up for dead."

In 1935, two schoolteachers, leading a large group of elementary students on an educational outing, decided to explore the catacombs and caverns beneath Hal Saflieni. According to witnesses, "as the last student turned the corner, the rope was clean cut." This "accident" was followed by a sudden "cave-in" which completely sealed the tunnel into which they'd gone.

While "caving" or spelunking accidents occur regularly, this event was preceded by another a few days earlier which put the entire matter into a different perspective. As related to Riley Crabb by a Miss Lois Jessup, and recounted as follows in "The Reality of the Underground:"

> *"What's down there?" she asked the guide, pointing to a small opening off the walls. "Go there at your own risk, and you won't go far," he replied. This was a challenge Lois couldn't pass up. She talked it over with her friends. Two of them decided to stay with Joe. The other three summoned up enough courage to explore with her. "I was wearing a dress with a long sash that day and as I decided to lead the group I asked the fellow behind me to hold onto it. So, with half-burnt candles in our hands the four of us started through that low, narrow passage, groping and laughing our way through. " I cam out first, of course, onto a ledge pathway only two feet wide, with a sheer drop of fifty feet or more on my right and the wall on my left.*
>
> *I took a step forward, keeping close to the rock wall side. The person behind me, still holding on to my sash, was still in the tunnel. "I held my candle higher and peered down into the abyss, thinking that with this dangerous drop it was better not to go on further without a guide. Then I saw about twenty persons of giant stature emerge from an opening deep below me. They were walking in single file along another narrow ledge down below. Their height I judged to be about twenty to*

> *twenty-five feet, since their heads came up about half way on the wall on the opposite side of the cave. They walked very slowly, taking long strides. Then they all stopped, turned and raised their heads in my direction. All simultaneously raised their arms and with their hands beckoned to me.*

The movement was something like snatching or feeling for something, as the palms of their hands were turned down." By this time her friends back in the passage were becoming impatient of the delay. There was a tug at the sash. "Go on. We're all getting stuck in here. What's the matter?" "Well," stammered Lois, "there's nothing much to see." She took another hesitant step forward, her candle in her right hand, her left hand against the cold rock for support. but it wasn't on a cold rock wall, it was on something damp and wet, AND IT MOVED! "Then a strong wind came from nowhere and blew my candle out! Now I really WAS scared in the darkness. I yelled to the others, "GO BACK! GO BACK! Guide me with my sash. I can't see!" "They pulled me back into the low tunnel and we backed up all the way along the passage into the larger room." Lois was relieved to see her friends and Joe, the guide, again. "Did you see anything?" one of them replied. "No, my candle went out," she replied with finality. "There was a strong draft in there." "Let's go," said Joe, looking at Lois, and she returned his glance eye for eye.

She knew beyond any shadow of a doubt that at one time Joe had also seen those giants. There was an expression of caution in his glance which held her to silence. "Out in the hot Malta sunshine again we thanked our guide and as we tipped him Joe said to me: "If YOU really are interested in exploring further it would be wise to join a group. There is a schoolteacher who is going to take a party exploring soon." Lois left her address with him, suggesting that he have the schoolteacher get in touch with her; but she never heard any more of it.

Some few days later one of the friends of the Hal Saflieni excursion called her on the phone. "Remember that tunnel you wanted to explore in the Hypogaeum?" Well, it says here in the local paper that a schoolmaster and thirty students went exploring and apparently got as far as we got. They were roped together, with the end of the rope tied to the opening of the cave. As the last student turned the corner where your candle blew out the rope was clean cut. None of the party was found because the walls caved in." Miss Jessup was shocked by the news, but it only strengthened her own resolve to say nothing of what she had seen and felt that unforgettable day in Hal Saflieni.

Some months later her sister came to Malta on a visit, and insisted on touring the famous Hypogaeum. Reluctantly, Lois went along, retracing the same route but this time with a different guide! She awaited that fateful opening with a dread expectancy as they worked their way through the corridors and rooms to the lowest level. The entrance to the tunnel was boarded up! So giant, hairy humanoids, described long before "sasquatch" and his various relatives became the center of tempest-in-theoretical-teapot debates among hominid postulators who ignore an underworld connection (largely out of realization that there own theories may have to be modified), were seen and described in detail by a frightened young woman. A few days later, a large group of children and at least two adults disappeared in the SAME cavern, forever. "Surely," the renowned cryptozoologist may pontificate, "if these creatures were actually seen, then they represent an unknown hominid which is wholly part of the natural (i.e., "surface") world. They can be handily categorized, fit into neat bestiaries of our own devising, and explained away. They are not seen over-running Malta in their full, towering, twenty-foot height, because they only come out at night and crawl around on their bellies in order to avoid detection."

According to Crabb, Miss Jessup described these subterranean beings in the following manner, her description embellished with Crabb's own interpretation of their nature: Actually, Lois found the giants of the cave hard to describe because of their covering which seemed to be like long white hair, combed downward and shaggy looking.

Their heads were unusually elongated at chin and top with large features, and the hair on their heads fell about the shoulders like a draped monk's cowl. Lois found the Heindel drawings exciting because "the currents in the desire body" sketches were the first to resemble in any way the cave dwellers she saw on Malta. Nor does her description of them correspond to Shaver's Deros, hideous dwarfs or trolls who might very well have carved that portion of Hal Saflieni now open to the public. This conflict in sizes and types very well illustrates the point I made earlier, that the underworld is peopled with beings of many sizes, shapes and varying degrees of density, from the completely physical to the completely invisible. Which leads to our next section.

(See Color Inserts)

3. THE PROBLEM OF DENSITY

Adjustable or varying density seems to be a trait or ability shared by apparitions, unknown humanoids, and some "cryptid" creatures. UFOnauts are said by witnesses to be able to alter their own physical density, pass through solid walls and doors, yet also become completely physical when they wish or need to be so. In 1987, a group of Russian cosmonauts aboard Salyut 7 reported to their mission commanders that they had been visited while in space by gigantic, nebulous beings, which they called angels. These beings, beside whom the cosmonauts and their vessel were but insects, smiled benignly, enigmatically, and perhaps somewhat amusedly at the tiny human interlopers into space. Vladimir Solevev, Oleg Atkov and Leonid Kizim went on record as saying, "What we saw were seven giant figures in the form of humans, but with wings and mist-like halos as in the classic depiction of angels." Of course, traditionally, angels can alter their density at will, becoming solid and visible, solid and invisible, or ethereal and visible, as the mood strikes them. When taking on a human form, as in Old Testament accounts, they are approximately of human size and apparent density. UFO entities are often described in similar terms.

According to Bill Moore and Charles Berlitz, the late paranormal researcher Gerald Light was invited to a top-secret meeting at Edwards Airforce Base in 1954. Here, apparently for the sake of eliciting an opinion from him, Light was shown a variety of captured "alien" aeroforms or craft, as well as some of the entities who were "instructing" the military in their operation. According to a letter from Light, written to researcher Meade Layne of Borderland Sciences Research Institute (and reported by Berlitz and Moore in their book, The Roswell Incident), these beings were capable of altering their physical density at will, as well as their visibility. He referred to them as "etherians." They could appear as solid and "real" as a human being, or "fade away" by degrees. One section of the letter expresses Light's horror or shock at seeing these beings for himself, and the effect of their reality on the usually unflappable military men around him:

> *To watch strong minds flinch before totally irreconcilable aspects of "science" is not a pleasant thing. I had forgotten how commonplace such things as the "dematerialization" of solid objects had become to my own mind.*
>
> *The coming and going of an etheric, or spirit, body has been so familiar to me these many years I had just forgotten that such a manifestation could snap the mental balance of a man not so conditioned. I shall never forget those forty-eight hours at Muroc!*

Previously in the letter, Light refers to these "other" humanoids in the following manner: "During my two days visit I saw five separate and distinct types of aircraft being studied and handled by our air-force officials-with the assistance and permission of the Etherians! I have no words to express my reactions." The key term here is "Etherians," but the ability to manifest in a "solid" physical form does not go unnoticed. Many entities, from the recent evil dwarf in Argentina called "El Diablito," to El Chupacabras, the "Hopkinsville Goblins," the Jersey Devil, and a wide variety of hairy humanoids of many sizes and descriptions, exhibit characteristics of extreme density, invulnerability to shotgun-blasts, bullets, shovels, vehicle impact, and other blows, and an unnatural strength or musculature which seems to enable abnormally high jumps, leaps, bounds, or speeds. This strength also enables them to rip into animal pens as if they are made of paper, or to escape any confines or snares which are set for them.

If examined logically, this variation of density becomes a clear characteristic of at least some anomalous entities, although not of all. If, as evidence overwhelmingly indicates, "El Diablito" and El Chupacabras, as well as many of the others mentioned, have a deep subterranean or "cavern-world" origin, then this increased density and strength may be due to an origin or ongoing habitation at pressures much greater than those which exist in the surface world. So where do the other, less "animal-like" beings, specifically angels and UFO entities, tie in? Unlike the others, they seem to be able to alter their physical density, vibratory level, or distance between their constituent atoms at will. This may very well be an attribute of beings of "spirit," but they are also physical when it suits them to be so.

Reflecting upon the giant angels as seen and sworn to by the cosmonauts, one has to note that, in an area of almost zero pressure and far from the Earth's most powerful electromagnetic and gravitational pulls, these beings were "greatly expanded" in size. Could it be that, upon reaching the surface of the planet, whether in a visible or invisible state, they would be naturally "compressed" to human size in some way, perhaps due to the electromagnetic and gravitational fields of the Earth? Conversely, is the super-hard or seemingly preternatural density, impervious nature, and excessive strength of more animalistic entities evidence of their subterranean origin at higher pressures, or nearer to a deep region of greater magnetic or gravitational force?

Some beings are apparently capable of altering their density at will, and these types seem to be initiators of encounters, the intentional interlopers and manipulators of human lives. The other category of entity seem to be types of purely physical creatures, yet evidence would seem to indicate that these beings may very well be in the service or at the beck-and-call of those with more "spiritual" density-altering abilities, and almost certainly of some sort of subterranean origin. The same can be said, perhaps, of other humanoid races or specie, particularly those who are interested in human and livestock genetics and flesh.

When anomalous entities are examined in the light of the foregoing theory, scientific reality and common sense seem to merge with the mystical or the outrÈ, to present an explanation which reconciles these things to our own physical reality. If this "variable density," or even organic densities different from our own are maintained, controlled, or otherwise affected by electromagnetic fields, or the EM field of the Earth itself, then this would explain much in terms of traditional beliefs about such beings. In many folk traditions around the world, iron is believed to have a particular efficacy against the influence of underworld beings. This is as true in Celtic and British folklore as it is in some Japanese and other accounts.

Much of this data is discussed in "Hidden Neighbors". After careful consideration of some of these accounts and beliefs, and taking into account many apparent electromagnetic connections which exist in regard to hauntings, UFO activity, ley lines, "places of power," and so forth, it seems that perhaps iron has a disruptive effect upon electrical or electromagnetic fields and the stability of some degrees of density which are utilized by the more "etheric" beings. When one considers that we are all, in effect, electromagnetic creatures, this makes some degree of sense.

The Highland Scots, Gaelic Irish, and the Norse were especially aware of this connection, and it's mentioned by Robert Kirk in The Secret Commonwealth, wherein which he states that in fact not only iron, but some "ultimate lodestone" in some depth of the north is most greatly feared by the "subterraneans." Does the electromagnetic field of the Earth present an underlying explanation or connection which provides a logical "unifying field" for things like telepathy, precognition, clairvoyance, reincarnation, and so on? We know that the human brain is a radio transmitter and receiver which is also highly electromagnetic in nature and activity; the most powerful naturally-occurring electromagnetic field that we know of, and which we all bathe in constantly, is that of the Earth itself. Some "hauntings" are thought to be imprints or recordings of extremely emotional events in the environment, and this is an electromagnetic phenomenon which has often been measured with electronic equipment. Perhaps "ghosts," perceived "memories" of reincarnation, and "insight" or clairvoyance are all recordings, "stored data," or accessing of information which is naturally held in the electromagnetic field ("akashic record") of the Earth.

By the same token, "places of power," mystical locations, ley lines, vortices, and so forth all are distinguished by a much higher-than-normal electromagnetic field. This concept doesn't negate any clairvoyant knowledge, experience or even a reincarnation memory in any way, but perhaps people who are of such a sensitive, "receiver" nature can consciously or subconsciously "load the data," which is stored somehow in the electromagnetic-physical matrix of the planet, as their own and replay it nearly perfectly. This may be due to something distinctive to their brain structure, and may part of the reason that UFOnauts, fairies, elves, succubi, incubi, and other beings exhibit interest in specific racial or cultural groups. Often these groups are those who exhibit higher-than-average extrasensory or precognitive abilities, which the Highland Gaels of Scotland called "the Sight." There seems little doubt that "things of spirit," and of some things which are connected to "spiritual" mysteries and anomalous entities, exhibit some sort of electromagnetic connection. The "disruptive" power of iron on subterranean "fairies," the alleged interest of modern UFOnauts or "reptoids" for sources of natural magnetism or high-voltage power-lines, and other mysteries may be brought into a more logical perspective if we view ourselves all as being rather like programs and subroutines which "run" on the electromagnetic (multi-dimensional) and physical hard-drive of a spinning dynamo that we call the Earth.

This perception of other regions on the electromagnetic spectrum may also explain some aspects of abduction experiences, out-of-body experiences, and "astral" travel. Lucid dreams which take on an "astral" tone may also be due to one's brain tapping into another spectrum of reality which exists on, or possibly within, our planet. George Wagner of Cincinnati, Ohio, relates the following "dream":

> *Although I've not had an "old hag" experience, I HAVE had a couple or three nightmares that went FAR beyond the usual breed. These experiences were the most terribly frightening I've EVER had, asleep or awake. It was as though I had plugged myself directly into the electric light circuit of evil. A friend asked me how I'd compare the evil of these dreams with, say, an extremely frightening horror movie. I answered that you'd have to multiply by many thousands. '"Look! He SEES the Demon-Bishop!" In the worst of these dreams, I found myself led by people I knew and trusted into a large, almost cavernous room.*
>
> *A huge and monstrous infant sat in the center of this chamber, an immensely fat baby weighing many tons and at least six feet high even in its seated position. It was naked except for an incredibly filthy diaper and a soiled and stained bishop's miter on its head. The infant held a bishop's crosier. The end of this staff was a human skull which seemed to be alive. The eyes glowed red and the jaw nattered and chittered all sorts of soundless blasphemies. There were also tiny, "living" skulls all over the miter on the baby's head. The crowd which surrounded me worshipped this monstrosity. Huge black flies crawled over the baby's body and circled in festering love-orbits around it. It was STRONGLY impressed upon me that if I too worshipped the baby, I could also become one of these dreadful black flies. [Hey, what a deal!] (This whole tableau struck me as violently as a sledge hammer blow. Waves of evil rushed at me, gushed at me, like lava from a volcano. And the whole scene was filled with a tremendous BUZZING sound.*
>
> *Ever since this dream I've been certain of one thing: evil BUZZES.) Now the evil became truly unbearable. I KNEW that the God of the Bible, of Jacob and Moses, of Mark and Luke and Paul, was a most impious myth. The Demon-Bishop was indeed God and I must go down*

on my knees in abject adoration. One knee touched the ground and the other was less than an inch from the surface. I knew that in a second or two I would be damned forever. So I tried to call out to Jesus Christ. But my tongue seemed to fill my entire mouth, so that I could manage nothing more than a croaking "Yedus Ki! Yedus Ki!" '"Listen to him!" the crowd cried out. "He still believes in that moth-eaten old fairy tale!" But "Yedus Ki" heard my prayer and I awoke in bed. My bed seemed to be rolling up in down, like a rowboat passing through rough breakers.

I immediately leaped out of bed, and the entire room seemed to be lurching up and down, like a crazy elevator (I assume this was all an inner-ear disturbance.) My mattress was SOAKED with sweat. I even squeezed drops of water out of it. But I then flopped back into bed, fell immediately asleep, and slept like a baby until morning.

The "buzzing" sound which Mr. Wagner relates as being an integral part of his experience, he has also felt at other times when something "unusual" seemed to be occurring. He's not alone; UFO abductions and encounters, out-of-body experiences, "old hag" or "succubus" attacks, and many other strange phenomena are accompanied by sensations of buzzing, vibrating, or even powerful or paralyzing electrical and vibratory sensations. Did George Wagner leave his body, as he seems to think he might have done, and was his consciousness or personality "somewhere else," beneath the ground?

Folkore or religious connections jump out here. First, one of the chieftains of the fallen Seraphim (and perhaps one of the Nephilim or Anunna/Anunnaki) is the demon Beelzebub (Hebrew "Ba'al zebub," or "Ba'al, Lord of the Flies"), who has the traditional title of "LORD OF THE FLIES." Second, Beelzebub is also the demon of gluttony: eating, drinking, and sexual excess. He's often portrayed as either a giant fly, or as an obese, huge creature. This sounds precisely like the thing in George's dream. The 'obese' baby of huge, gluttonous size, controller of all the "flies" who were damned or captured human souls, seems likely to have represented or been no one other than the personality who is traditionally known as Beelzebub. The "bishop" reference or title would be indicative of a “hierarchy of mockery” which is in place, in which the structure of the Catholic Church is mocked.

The demonological connections of the symbolism simply fit together too perfectly to be coincidence. George seems to think that this experience has something to do with his proximity to a “window region” for strange phenomena, UFO sightings, unknown creatures, and paranormal activity, the region where Ohio, and heavily-caverned Kentucky and West Virginia come together. Perhaps not coincidentally, this region is also the haunt of “Mothman,” the “Hopkinsville Goblins,” and other strange beings.

The “disoriented” motion or sensation experienced by George upon awakening is also not unfamiliar to others who have experienced an “altered state” of consciousness, and been abruptly returned to the body. This may imply that the electromagnetic component of the mind which is the “personality,” the “soul,” or something comparable, is out of alignment with the physical brain when first “pulled back into place.” And underlying the entire account is the “buzzing” electromagnetic sensation, a clue at the very least to the nature of the phenomenon, and perhaps a clue to the layered realities of varying densities which are beneath, above, and around us, and always have been. Perhaps “ET” should stand for “Electromagnetically Transmutable,” and an “imp” should be called an EMp.

The 20th Century saw a new spin on the "Giants in the Earth:" The "Shaver Mystery."

(See Color Inserts)

Part VII: The Nephilim Giants

Who were the Nephilim? The heart of the matter lies in the ancient accounts of the Nephilim and their offspring–who and what they really were, what they did, where they are now, what their agenda is, and so on.

Conclusions about the nature of the Nephilim, as borne out by quotes from the book of Enoch, as well as the Books of Jude and Revelation, present a remarkable picture, as do the actual mentions in the original Old Testament of the Nephilim, Anakim (Anunna/Anunnaki), Rephaim, and Rapha–along with others descended from Nephilim forbears. In fact, the traditional Hebrew view is that the "rapha," meaning, interchangeably, "irrevocably dead," "evil spirits," "demons," and "giants," are the spirits of the half-angelic Nephilim offspring that were killed in the great cataclysm we remember as the Flood. This is not a tall tale, but a truth of which Christ and his disciples were aware. This is why they quoted from the book of Enoch.

The story can also be found in the books of Jasher, Jubilees, and the recently-found Dead Sea Scroll, the Book of Giants. Angels have bodies. They can take a variety of forms, as well ("for Satan can appear even as an angel of light," etc.). There were two angelic rebellions, the first being the Luciferian rebellion that most are familiar with; this is the source of what

Paul calls, in the sixth chapter of Ephesians, "Spirit wickedness in high places." This is why Jesus calls Satan "the prince of the powers of the air."

UFO researchers, take notice. The second rebellion came much, much later, and is well known in Hebrew esoteric tradition. In fact, much of this knowledge is still intact in the Hebrew tradition. Many books which were originally in the Bible–the books of Enoch, Jasher, Jubilees, and so on–contain the whole story of the fallen angels who rebelled in order to mate with human women. These beings were called the NEPHILIM, literally "those who came/fell down."

A recent dead-sea scroll find and translation, "the Book of Giants," also contains this story in great detail. Before you question the validity of the non-canonical books I mention, keep in mind that not only were they part of the original Old Testament, but many of them–particularly Enoch–were quoted from by the Apostles in the New Testament. Quotes from the book of Enoch can found in the books of Jude, Revelation, and several other places. The book of Enoch pre-dated the time of Christ by around 300 years. In the Old Testament, Genesis tells us that "the sons of God looked upon the daughters of men and found them fair, and took of them wives as they chose." The result of these unholy, unsanctioned unions were "the mighty men of old, the men of renown." The actual translation of this passage reads something like this: "The nephilim came and took human wives for themselves, as many as they wanted. Their children were the nephilim of old, the giants."

The books of Jasher, Jubilees, Enoch, and others contain the complete story of what happened in the antediluvian (pre-flood) world. The Nephilim angels set themselves up as pagan "gods" to be worshipped, and their hybrid, genetically monstrous sons as "demigod" kings over the rest of humanity. These offspring were also called nephilim, and they were brutal, cruel, occultic and cannibalistic giants who treated ordinary human beings with contempt. There were 200 original "Watchers" above the Earth, who rebelled and became the Nephilim. Their leader was named Semyaza. Through deceitful angelic knowledge, the pre-flood world was corrupted.

The book of Enoch says that the Nephilim and their children "sinned against all flesh," creating hybrid creatures which were part human, part beast, part fowl. Sound familiar? The 200 fallen "watchers," whose original job had been to PROTECT humanity and the Earth, were utterly depraved. They taught mankind the arts of witchcraft, sorcery, drug abuse, and war. The gene-pool of the human race was tainted as well, almost beyond recovery. When you read the story of Noah and his family, it should be understood that he was not only selected to survive because of the quality of his character; Genesis states that, unlike the majority of the rest of humanity, Noah was "found perfect in all his generations." In other words, he and his family were untainted by the fallen angelic bloodline, which God had cursed to eternal damnation (in the Hebrew Old Testament, the word "rapha" means "the damned," the "eternally dead," "demon," "ghost," and giant, i.e., the spirit of one of the offspring of the Nephilim.

After the flood, when most, but not all, of the Nephilim offspring and hybrid mutations were destroyed, the 200 former "watchers" were confined inside the Earth, in a prison from which they can not escape. These are those who are referred to by Paul in Ephesians chapter 6 as "the rulers of the spiritual darkness of this world." Although imprisoned, they are still VERY powerful and can influence world events, particularly through their offspring and creations. Jude refers to them, quoting directly from the book of Enoch, when he says "the angels who are chained in eternal darkness, awaiting the judgement of the great day." This is NOT a reference to Lucifer/Satan and his angels, who still "wage war in heaven." The more one learns about the ancient knowledge and traditions, the more it is seen that a spiritual yet physical and galactic war is behind all the myths, legends, religions and traditions of our world.

This a very complex topic, but the information is out there if you want to look for it.... Regarding this passage from Genesis chapter 6:1

> *And it came to pass, when men began to multiply on the face of the earth, and daughters were born unto them, 2 That the sons of God saw the daughters of men that they were fair; and they took them wives of all which they chose. 3 And the LORD said, My spirit shall not always strive with man, for that he also is flesh: yet his days shall be an hundred and twenty years. 4 There were giants [nephilim] in the earth in those days; and also after that, when the sons of God came in unto the daughters of*

men, and they bare children to them, the same became mighty men which were of old, men of renown. 5 And God saw that the wickedness of man was great in the earth, and that every imagination of the thoughts of his heart was only evil continually.

The passage "and also after that" refers to a time after the flood, when an unspecified form of Nephilim or fallen angels once again took an interest in human women. These are not the same as those nephilim who did this before the flood however, as they are imprisoned. The result of these later unions were the Canaanite and other "giants" of the Old Testament, such as those mentioned in the Book of Numbers and elsewhere: "`And there we saw the Nephilim [the sons of Anak, who come from the Nephilim]; and we seemed to ourselves like grasshoppers, and so we seemed to them.' Then all the congregation raised a loud cry; and the people wept that night." (Numbers Chapter 13 verse 33)

The Anakim or Children of Anak are synonymous with Zechariah Sitchin's Anunnaki, or Anunna. There was a specific reason that the Israelites were commanded to utterly wipe out these peoples, as this represented a new effort to undermine the planned destiny of humanity. The issue of technology, based on advanced scientific (fallen angelic) knowledge, comes into play in all of this as well. UFOs, particle beams, all types of advanced technology are described in both the Old and New Testament. This demonstrates that a physical presence is required by both sides in the ongoing conflict, and technology is required in order to function in our reality. UFOs relationship to predatory interlopers is made obvious in many places, particularly in the book of Zechariah.

Apparently a certain amount of this activity was allowed or permitted by God, among those who followed the ways of the Nephilim who had set themselves up to be worshipped and emulated:

1 Zch5:1Then again I lifted up mine eyes, and saw, and behold a flying scroll. 2 And he said unto me: 'What seest thou?' And I answered: 'I see a flying scroll; the length thereof is twenty cubits, and the breadth thereof ten cubits.' 3 Then said he unto me: 'This is the curse that goeth forth over the face of the whole land; for every one that stealeth shall be swept away on the one side like it; and every one that sweareth shall be

> *swept away on the other side like it. 4 I cause it to go forth, saith the LORD of hosts, and it shall enter into the house of the thief, and into the house of him that sweareth falsely by My name; and it shall abide in the midst of his house, and shall consume it with the timber thereof and the stones thereof.' 5 Then the angel that spoke with me went forth, and said unto me: 'Lift up now thine eyes, and see what is this that goeth forth.'*

In ancient times, a "scroll" was one rolled-up piece of leather or parchment, tightly bound. In other words, this was a cylindrical flying object of some sort, one which moved with intent and purpose. When one considers the "cybernetic" aspects of many unknown creatures and entities–metallic hardness, metallic sounds, luminous eyes, and so on–it becomes apparent that we are dealing with an advanced and ancient technology, one which easily melds and utilizes genetic material and inert technology.

These qualities have been noted in strange creatures for centuries, but recent beings with such qualities abound: Mothman, El Chupacabras, the Hopkinsville/Kelly Goblins, "El Diablito" (a super-dense, violent dwarf) in Argentina, and a variety of other creatures described as more or less impervious to machete or shovel-blows, pick-blows, vehicle impact, shotgun blasts, and handgun ammo. Again, all of these things tie together to give us an overall bigger picture. Similar traditions exist in the Hindu Vedas and Puranas, Norse mythology, Greek mythology, and elsewhere.

The Judeo-Christian tradition, in my opinion and after twenty-plus years of study, is the most accurate one in terms of painting a picture which may represent some sort of historical accuracy. The oldest written language which we know of is that of ancient Sumer (the biblical Shin'ar). Ur was a city in this region (later called Chaldea), and it was from Ur that Abram (later called Abraham) came. He was the son of a pagan priest. How do we know this? Because cuneiform tablets have been found which describe the leaving of Abram from Ur, along with his possessions, servants, family members, and so on! And these match the biblical account.

The oldest extant written story, tradition, or text that has been translated is the Epic of Gilgamesh, a half-human demigod who lamented his mortality, and who went on a journey (part of it below the ground, in the underworld) in search of immortality (like that of his father). An interesting observation: In the "Book of Giants" (4Q203, 1Q23, 2Q26, 4Q530-532, 6Q8), a Dead Sea scroll find, several giants, and their Nephilim fathers, are mentioned by name. One of them is Gilgamesh, the same Gilgamesh from the Sumerian tradition, a "mighty man" who is lamenting his own impending death and damnation:

> 4Q530 Col. 2 1 concerns the death of our souls [. . .] and all his comrades, [and Oh]ya told them what Gilgamesh said to him 2[. . .] and it was said [. . .] "concerning [. . .] the leader has cursed the potentates" Enoch sends a tablet which pronounces the sentence of the flood, to the giants (who were rulers over men): 4Q530 Frag. 2 The scribe [Enoch . . .] 2[. . .] 3a copy of the second tablet that [Epoch] se[nt . . .] 4in the very handwriting of Enoch the noted scribe [. . . In the name of God the great] 5and holy one, to Shemihaza and all [his companions . . .] 61et it be known to you that not [. . .] 7and the things you have done, and that your wives [. . .] 8they and their sons and the wives of [their sons . . .] 9by your licentiousness on the earth, and there has been upon you [. . . and the land is crying out] 10and complaining about you and the deeds of your children [. . .] 11the harm that you have done to it. [. . .] 12until Raphael arrives, behold, destruction [is coming, a great flood, and it will destroy all living things] 13and whatever is in the deserts and the seas. And the meaning of the matter [. . .] 14upon you for evil. But now, loosen the bonds bi[nding you to evil . . .] l5and pray.

The hybrid beings mentioned in other books such as Enoch and Jubilees are also in the Giants Scroll:

> 4Q531 Frag. 2 [. . .] they defiled [. . .] 2[. . . they begot] giants and monsters [. . .] 3[. . .] they begot, and, behold, all [the earth was corrupted . . .] 4[. . .] with its blood and by the hand of [. . .] 5[giant's] which did not suffice for them and [. . .] 6[. . .] and they were seeking to devour many [. . .] 7[. . .] 8[. . .] the monsters attacked it. 4Q532 Col. 2 Frags. 1 - 6 2[. . .] flesh [. . .] 3al[l . . .] monsters [. . .] will be [. . .] 4[. . .] they would arise [. . .] lacking in true knowledge [. . .] because [. . .] 5[. . .] the earth [grew corrupt . . .] mighty [. . .] 6[. . .] they were considering [. . .] 7[. . .] from the angels upon [. . .] 8[. . .] in the end it will perish and die [. . .] 9[. . .] they caused great corruption in the [earth . . .] [. . . this did not] suffice to [. . .] "they will be [. . .]

So the oldest written human tradition that we know of is reflected in the Hebrew tradition of the Nephilim and their hybrid children. The monumental task at hand is to deduce, if possible, where and how Mothman (identical to the Sumero-Babylonian UTUKKU, a demon of the KI GAL or underworld) and other anomalous entities fit into this tradition. Additionally, there's an entire related tradition of the "pre-Adamic" or "pre-Adamite" world, which was peopled by humanoid beings of various types, along with other strange beings. These may be some of the same hominids that are represented in the fossil record; and some of these forms, as survivors and anachronisms, may explain some sightings of anomalous beings.

Lilith may well have been of one of these races, and she fled into the wilderness, where she mated with subterranean "demons" and other beings, swearing that she and her children (the "Lilim") would always prey on human children, and seduce human beings into hybrid-producing sexual encounters. Again, this is paralleled in nearly EVERY ancient tradition, from the Lamias of Greece, to the Huldre-folk and seductresses of Scandinavia and the Dragon-kings, queens, princes and princesses of China and Japan.

Part VIII: The Shape-Shifting Myth and Mystery

One of the oldest types of entities in folkloric and mythological accounts is also one of the latest, and of late, more common in paranormal or fortean accounts. Various entities may, upon initial consideration of witness accounts, seem widely-disparate in nature; yet whether of "gray" alien, the original Puerto Rican "chupacabras," "grinning men" (of various sizes, heights and descriptions), MIBs, or "reptoid" type, all seem to share a variety of characteristics which can be more or less identifiable as "reptilian" or "amphibian" in nature. These characteristics range from "leathery" or lizard-like skin texture, bulging eyes, slit or oblique pupils, webbed hands or fingers, and to some extent, sparseness or absence of facial and body hair.

The reptilians or "reptoids," stretching back through the centuries, have had a folkloric reputation of being deceivers, shapeshifters, vampires, and anthropophagous. Sightings and encounters have provided the basis for legends of horned, leathery demons and imps; goblins; ghuls; "henkies" and other dwarfish, somewhat-humanoid forms, nagas, trolls, and so on. The interest in human blood and genetic materials, exhibited by fairy lovers and abductors, elves, trolls, undines, incubi, succubi, frog princes, jinn, "deros," and the elemental or sylph-like "gray aliens" is all too obvious.

Of particular interest is the ability, currently attributed to the "reptoids" or lizard-men of ufological and conspiracy accounts, to "shape-shift" or take on an illusion of humanity. Some accounts have the reality the other way, in which the human being, due to stimulation of an non-human bloodlust or other desire, is transformed or "shifted" into a reptilian humanoid. While the latter type of "sighting," when genuine, may indicate a form of demonic (spiritually "reptilian") possession or oppression which is so powerful as to be visible for a temporary time in the visible spectrum, it is not necessarily the same as the encounters reported, through the centuries, of "real" shape-shifters, reptilians, devils, fairies, and "aliens" which can take on a deceptive or even a familiar form, for purposes of predation.

The ability of the fairies or "good neighbors" (a euphemism) to use "glamour," of nagas and rakshasas to utilize sorcerous illusion, and "vampires" and their "hypnotic" and shape-changing abilities, are reflected, in modern variants, in the reports of shape-shifting reptoid imposters, alien abductors who appear as familiar friends or family members, or the "rays" generated by dero "mech" which create a holographic persona. What is the "true nature" of these beings, and what is their common link, if any?

In the book "Etidorpha," a man is taken on a tour of the underworld and cavern realms by a guide who is more or less humanoid in form, but is described as having skin or flesh of an amphibian-like texture and no body hair whatsoever. This being is also compared to a lizard, and is described as being "slimy," "cold and clammy," and so on. The "guide" has one distinguishing characteristic which stands out in particular, however: "he" has no face at all, just a smooth ovoid patch where a face should be. As it turns out, he is not alone.

"Faceless" entities have long been present in apparition, ghost, and monster stories. Could these forms have any relation to the "shape-shifting" types of beings described? Are they not just reptilians, but representatives of a type of life which is both parasitic and chameleon-like? The question which comes forward is this: Could a "faceless" form of "genome-thief" still not be essentially reptilian, or a type of mammalian/reptilian creature descended from reptiles (as we humans, and all mammals, supposedly are)? Chameleons alter their skins cells to reflect colors, patterns, and even moods like aggression or passivity; could there be a form of life which has evolved this to an ultimate form of mimicry or deception, actually using stolen or absorbed genetic material? Would this possibility provide an answer to MANY mysteries?

There must be a clue, an explanation as to the "malleability" of these beings, their transformational abilities or else powers of genetic manipulation. Perhaps some of them are like "blank genetic templates," upon which an entire genome can be temporarily or permanently inprinted; or even different types of genomes, stolen from other life-forms, are imprinted or combined. Here would be an answer to the chimaerical and "hybrid" forms of many of the more outrageous "goblins," "cryptids" and ufonauts!

In an October 20, 1991 lecture, Ufologist Michael Lindemann delivered a lecture which contained this first-person witness account of an encounter in a top-secret underground military site:

> *My friend who worked in the underground bases, who was doing sheet-rock was down on, he thinks, approximately the 30th level underground... these bases are perhaps 30-35 stories deep. As I say they are not just mine shafts, these are huge, giant facilities... many city blocks in circumference, able to house tens of thousands of people. One of them, the YANO Facility [we're told... by the county fire dept. director, the county fire dept. chief who had to go in there to look at a minor fire infraction] there's a 400-car parking lot on the 1st level of the YANO Facility, but cars never come in and out, those are the cars that they use INSIDE.*
>
> *O.K., so... a very interesting situation down there. Our guy was doing sheet-rock on the 30th floor, maybe the 30th floor, underground. He and his crew are working on a wall and right over here is an elevator door. The elevator door opens and, a kind of reflex action you look, and he saw three 'guys'. Two of them, human engineers that he's seen before. And between them a 'guy' that stood about 8 to 8 1/2 feet tall. Grey skin, extra-long arms, wearing a lab coat, holding a clip-board, AND HE DIDN"T HAVE A FACE! My friend said it was just teeth right there. Just teeth where a face should be...*

The Maya called such underworld shapeshifters "vhujunka." Other than the face, the physical description sounds much like that of one of the so-called reptoids reported at the alleged Dulce complex. Right down to the height, white lab-coat, and clipboard.

Are "reptilian" faces as much an illusion as anything else might be? Does this "faceless" account indicate that we're dealing with something that can appear in any form, or with any face, that it wishes? Underworld beings are traditionally the masters of "glamourie," or glamour, illusion, shape-shifting, deception, and so on, and of course, "Satan can appear even as an angel of light." But their interest in US is so self-serving and manipulative, yet tinged with desperation, that other questions logically follow.

Is it possible that we might be dealing with something which is *purely a genetic thief or parasite?* A thief of genomes, in whole or in part as the necessity arises, a "blank slate" predator which can assume a variety of appearances? As noted previously and repetitively, this has traditionally been the province of "underworld" or subterranean beings in folk traditions worldwide, from western fairies and demons, middle eastern jinn and dubuks, to Indian nagas and Japanese "foxes" or fox-people.

These "others" may be genetically diverse beyond our imagining, due to this ability, yet incapable of reproducing of their own accord, or maintaining a stable and comfortable face, form, or life-span without it....and how does this all tie in with both livestock and human mutilations, which are essentially identical in nature and tend to concentrate on the blood, mucous membranes, reproductive organs and surrounding tissues, eyes, "meaty" organs like the liver (which is highly regenerative), and so on? Could these materials be of particular importance to an advanced genome-imprinting or cloning technology, which utilizes genetic material that has been purloined from somewhere or someone else?

Perhaps they have perfected a form of tissue regeneration and repair which requires some of the same materials which have always been offered up in sacrifices to the dark, bloody "gods" of antiquity, and *this* is the secret to "their" longevity (remember, vampires "live" forever as long as they stay out of the sun, and sun-dodging fairies are generally considered "undying" as well). Gruesome photos of a human mutilation victim from Brazil, along with autopsy data, support yet again that the vital organs, along with the most successfully regenerative tissues–the blood, mucous membranes, glands, "sweet bread" organs (the often self-regenerative liver, as well as stomach, intestines, etc.), sexual organs, and so on–are always the focus of their selfish and brutal desires.

It seems at times that Keel, Steiger, and other researchers dance around this possibility and even hint at it, but never really come out and say it. Until now, it may simply have been too unspeakable to contemplate or dare to put into print.

Even if unseen manipulators and predators have a reptilian origin that's lost in a vast antiquity, could it be that they have a partially hominid, somewhat mammalian heritage as well? Could they be as closely related to humanity as the monotremes (egg-laying mammals) are to the vivverids, for instance? Are they an ancient form of predator which is in fact a parasite, who has evolved along with all hominids and culminating with a dependence upon modern humanity? Vaguely, distantly, related in the vertebrate sense, perhaps more reptilian than we mammals, yet so old a species or variety of specie that they have evolved beyond our labels, and now represent a third form which is divergent, to some degree, from both the reptilian and mammalian kingdoms; this indeed seems more plausible as time goes by.

We may need to re-evaluate the nature of our ancient enemies.

(See Color Inserts)

Part IX:
A Cave Fable

When one examines the possibility of mostly-unseen co-inhabitants of our planet, it is often met with skeptical scoffing, coupled with a complete disregard for the evidence which is in plain sight. The evidence is multitudinous in example.

Is there evidence for a subterranean or subterrestrial influence on human history, world events, and the fates of nations and human beings? Of course. One case in particular has had tremendous impact on human history, and probably on the eventual fate of our species in terms of whether or not we even survive, or destroy ourselves in a never-ending conflict.

Once upon a time there was a primitive desert tribesman of a bandit tribe, living in a very inhospitable part of the world. This young man was ostracized to some degree by his own people, who were an animistic, multiple-god-worshipping lot, warring continually among themselves for plunder and tribute.

One night this young man had a seemingly "mystical" experience. Alone in a cave, seeking shelter and sleep, he was awakened by a "luminous being" who began to proclaim to him pronouncements of a grandiose and disturbing nature–so disturbing, in fact, that at first the young tribesman believed that the being was "Shaitan" or an ifrit (a form of underworld djinni). This being spoke in terms of conquest and judgment, the end of the world, and the wrath of the god of the young man's tribe, who was in fact at that time but one among many pagan deities in the region, the god of the moon. His symbol was the upturned crescent moon, making "him" a variant of the Horned God (and this symbol is still in use to this day). The entity declared that it was the messenger of this god, and called for the utter destruction and abolition of all other gods, deities, and belief-systems – including the 360 other pagan deities which were worshipped by the tribesman's race at that time.

As a result of his encounter with the anomalous entity, the young man at first believed that he had been contacted by a demon of the underworld, out of the depths of the cave and the earth–but his ambitious wife and relatives soon changed his mind, telling him that he had been "chosen." As a result of this, the young man embarked on a brutal and murderous military campaign of savage butchery, theft, rape, and other atrocities, with bandits and outlaws as his initial army recruits. Over the course of time his own cruel personality traits were inserted into the supposed revelations he had received in the cave, and his proclivities for beheading, lack of mercy and tolerance, rapine, pillaging and theft, slave-taking, and pedophilia were not only accepted by his subjects, but were glorified within the tenets of the new "religion."

Now, over the centuries since this encounter with an unknown humanoid being in a cave, the world has seen millions of innocent Christians, Jews, Zoroastrians, Buddhists, Hindus, and others invaded, enslaved, raped, robbed, forced into conversion at the point of a sword, villified, degraded, and downtrodden by the adherents of the CULT which was founded by a non-human entity and its ambitious, credulous follower and high-priest. Millions of Christians and Jews, having lived peacefully together in ALL Middle Eastern countries for centuries, were invaded by waves of murderous fanatics from the desert, their cultures destroyed, obliterated, and looted. This plague spread to Asia, and at one time even well into Europe, to hammer on the walls of Vienna.

This religion originated in a cave with a nonhuman entity. During the "chosen one's" life, he had a number of strange experiences, including one in which the same "entity" from the cave came to him and cut him open, took all of his insides out, and then put them back inside (or so he said, and so the sacred traditions of his cult still claim and celebrate). This is identical to some extreme UFO cases, and also to the activity of the entity that Cabeza de Vaca called "Senõr Malacosa" ("Mr. Badthing"), which plagued the Avavares Indians. Senõr Malacosa told the Avavares that he came from within the Earth or underground. The "dismemberment" of abductees is seen in many "extreme" abduction accounts, and is also a prevalent, common theme in many initiation-style shamanic experiences which allegedly take place in another reality, or altered state of consciousness.

The "chosen one" in the fable is renowned to this day for the "fits" that he would have, complete with seizures and babbling, which would be followed by new "accounts" of experiences only he had witnessed, along with attendant new "revelations".

Meanwhile, a large segment of a certain religion is waiting for a Messianic Warlord to return from *out of a well*, into which he/it descended 12 centuries ago, and from which he (or it?) will soon ascend, to literally lead them to victory and kill everyone else on Earth. The leader of a certain country which wants nukes has stated that he expects the return of this religious leader very soon, and knows that his nation will lead the others who practice this "religion" to victory over the rest of the world....

Scary stuff, indeed, and not something to laugh about.....And it involves quite a bit of "coincidental" subterranean material (there's even more if one digs deeply enough). And when an ultimate affront against the rest of humanity calls for an ultimate response, who will really be to blame: the deceivers or the deceived? Perhaps both....

Today this Cult of the Underworld, filled with inconsistencies, self-delusion, circular reasoning, self-contradictions, and hatred for all people of different belief systems, is still with us–and all because of a naive and impressionable barbarian tribesman's encounter with a deceptive and manipulative non-human being, in the depths of a cave.

End of Fable? Probably not.

Part X:
THE DECEPTION of the DEPTHS:
Humanity and the Ocean of Secrecy

(See Color Inserts)

1. OUR MYSTERIOUS HOME

Imagine that you have a house--a large and spacious house, perhaps a Victorian mansion. You have lived there for your entire life, yet the house is so large, that you and your family have more or less confined yourselves to the ten or twelve rooms you tend to use, and the rest of the house--at least three times as many rooms, if not more--is closed up and unoccupied. Neither you nor other members of your family care to deal with cleaning and maintaining the other areas, so doors are kept shut, windows closed, and furniture is covered with sheets and plastic. The space you all use suits you well; anything additional is extra work, so you let it lie. The "other" part of the house can be quite creepy at times, but the basement, especially, is deep and largely unexplored.

Upon occasion you hear inexplicable, mysterious sounds from the "closed" and unoccupied areas of the house: muffled voices and sudden unexplained bangings and thumping. Sometimes as you pass a hallway which leads to those areas, you glimpse doors closing suddenly, or a quick shape gliding across a darkened hallway to disappear into an "uninhabited" room. Things scamper to hide when you arrive unexpectedly, and you only catch the slightest impression of what they seem to be; and upon occasion you and the others in your family feel that, when you least expect it, something or someone creeps into your living areas, absconds with some of your belongings, spies on you in your most intimate or private moments, and invades your pantry and refrigerator with impunity. Beef and other meats seem to be a favorite item for theft. Occasionally you feel that you are being victimized as you sleep, or when you are mostly asleep, or in some sort of "altered state" which you don't understand. The most maddening aspect of all of this turns out to be that, in spite of the certainty that you and your family feel as to "something" going on, you can never catch anyone or anything in the act, nor can you prove that these invaders or squatters exist. Over time, you come to feel that you are either all insane, or that the beings are essentially benevolent. You condition your children to have little or no fear of the strange and mysterious areas and phenomena. This is more psychologically comforting than to admit that the relationship between your family and these "others" is identical to that which a host or prey population has with a predatory or parasitical species.

Now, expand the scope and imagine that your huge and spacious mansion is in fact a planet--our planet, the Earth, three-quarters of which is covered with seas, oceans, and large lakes, and much of which is inhospitable to human habitation due to extremes of temperature or other environmental, terrain, or subterranean conditions. You know very little about those unknown and unexplored areas, and what you do know you only learn second-, third-, and fourth-hand, and you have naturally been conditioned to believe what you are told. Yet those areas are apparently home to a myriad of strange beings and occurrences, the types of beings and occurrences of which folklore, myth, and legend are made. We humans regularly have strange and unsettling glimpses and encounters of these "others" and their parasitical, predacious activity, and we always have. Seemingly we deal with a superior technology or an alternate state of being which provides for invasive and often nearly-invisible encroachments into our areas of the "house." We take it all in stride, though, adding each event or encounter to our pile of human family lore, while the more fearful and disconcerted among our kind hide beneath a leathery shell of skepticism, poisoning their barbs against the unknown with ridicule and circular reasoning which ignores entire aspects of reality, along with tons of circumstantial, eye-witness, cultural, and anecdotal evidence.

(See Color Inserts)

This is the way things are.

This is the Truth.

Many people throughout history have had some glimpse of this "underlying reality." The fragmented picture presented by such brief peeks into the unknown is not heartening for humanity. UFO abduction accounts are absolutely filled with stories of people who sincerely believe that they were taken, usually against their wills, into vast cavernous spaces beneath the crust of our planet, or beneath the oceans. Of all modern UFOlogical accounts, these are perhaps the most-scoffed-at and ridiculed. Another physical reality here on our own, cozy, well-known little home of a planet? Impossible! Absurd! But these stories are nothing new, historically.

Some stories tell of women who are summoned as midwives, to attend fairy births or nurse sick hybrids. Often, however, a husband or lover would have to free them from the underworld realm

Celtic, Norse, Chinese, Japanese, and Hindu accounts all speak of subterranean and sub-oceanic realms which are inhabited by non-human, potentially hostile beings. The Irish sea-god Mannanan mac Lir and his particular race of "fairies" or gods lived in the twilight undersea realm of Tir na Nog, complete with what sounds like a sealed entry or portal into an underground region; while the Bhagavat Purana, Canto Nine, Canto 3, Texts 28 - 29, describes an undersea, subterranean kingdom called "Kushasthali," located "in the depths of the ocean." Some of the Hindu Nagas (serpent-people) were said to have their dwellings beneath lakes, rivers, and oceans, as well as being subterranean in general; and all the beings of these regions were fascinated with one type of surface-world creature, above all others: human beings. Abduction for purposes of marriage, rape, cruelty, even feeding purposes, are common themes throughout most mythologies, with some aspects only thinly-veiled at best. Children, virile youths, and fertile women are almost invariably of greatest interest to these "mythical" races.

In the fairy traditions of the Christian West, there are numerous accounts of "physically-real" fairies--usually in a watery environment--being observed and overheard as they lamented their own "lost" spiritual state. One question they seem to be greatly concerned with is whether or not they can be "reconciled with God," or "What will happen to me/us on the Day of Judgment?" According to these folklore accounts, the percipient responds in either of two ways: He or she states "God may afford you a pardon on that day, based on your actions throughout the centuries," or "You will never be part of the Kingdom of God." The first statement always meets with great joy and delight, before the entity disappears into the water. The second meets with the wails of a tormented soul, hopelessly lost.

Hans, half-human and half merman, comes into conflict with a cousin of sorts who has been terrorizing passersby. From "The Mermaid's Son."

Illustration: H.J. Ford

Indeed, folk and fairytales, religious accounts from various traditions, and even historical accounts of "inexplicable" disappearances bear this out, from the subterranean "Pied Piper" (who led the children of Hamelin into a "closing cave," never to return) and "changeling" tales of medieval and ancient Europe, to the abductions and rapes of women in ancient mythological traditions, by beings other than human. Surely, the modern scoffer would retort, this paranoid worldview is not borne out by modern science and knowledge? Such deluded ramblings are easily dispelled--or are they? Statistics from the National Center for the Victims of Crime (http://www.nvc.org/infolink/INFO30.HTM) might indicate that this type of predation may be ongoing, and in fact the numbers show that something strange is going on, and it is of a staggering extent:

"In 1998, 932,190 missing persons (adults and juveniles) were reported to the police and entered into the FBI's National Crime Information Center. The FBI estimates that 85% - 90% of missing persons are juveniles; therefore, approximately 800,000 cases of missing children (or 2,200 children per day) were reported in 1998 (National Center for Missing and Exploited Children, 1999)."

So, effectively, 2,200 children per day, on average, disappear in the United States alone. Almost a million children simply VANISH on an annual basis. Where are they all going? They're not all runaways, that much is obvious; in fact, as little as 8-10 percent of them are probably runaways. If the victims of serial-killing madmen, their bodies would be found by the thousands. Some sort of bizarre cultic connection may exist, but in the service of whom? This would still not account for the stunning number. Obviously, something unexplained is happening to these kids. They haven't vanished "upward," as a mass abduction of this type would have to be witnessed, so if they aren't on the dry surface of our planet, roaming the streets en masse, or vanishing into the sky, then logic would indicate that only one direction is left to which we should turn our attention—downward.

Why is there no huge media coverage of this horrific statistic? If this number of people were taken by war, disease, or violent crime, the public outcry would be deafening and the media would lead the charge; but isn't it strange that these huge numbers are not publicized or openly displayed in a meaningful way? For we ignore the "strangeness," the "violations" which occur within our own "mansion," conditioned as we are to turn off our critical observation (lest we be ridiculed, or labeled paranoid, strange, or insane) in the face of unspoken and undeniable horror--for this ongoing atrocity is equivalent to either genocide, or to... the harvesting of a crop.

And this set of statistics is that of only one nation, for one year. This is evidence, indeed, that something absolutely inexplicable, unnatural (by our definition) and sinister is indeed ongoing. But is there other evidence, evidence which may prove or strongly indicate that our children, livestock, and other abducted resources are vanishing beneath the oceans and crust of our planet? Again, ancient and modern accounts parallel one another.

The Pied Piper of Hamelin, leading the children into the underworld.

Illustration: H.J. Ford

Celtic and Scandinavian folk beliefs are particularly filled with accounts of fairies, trolls, huldre ("hidden folk") and other subterranean creatures who abduct cattle, dogs, horses and other livestock, or else cross-breed their own livestock with the "hardier" stock of humanity's domesticated species. There are also numerous accounts of "fairy cattle" ("Chrodh Mara") and "water horses" ("Cabyll-Ushtey" on the Isle of Man, "Each Usige" to the Highlanders) or "kelpies," which make their homes in lochs and rivers. These strange beasts venture forth to mate with their surface-world counterparts, or, as in the case of the kelpie, which looks like a horse in some respects but has a carnivorous appetite, in order to also attack and take a bite out of livestock, or out of a human being or two. Similar stories can be found in folktales from around the world. In terms of the origin of either the "dainty" or huge and fierce cattle and other creatures in such tales, the Gaels, Scots, Britons, and other peoples had no doubts: They were "underworld" or fairy creatures, descended from or distantly related to terrestrial forms.

The parallels between fairies who abduct and mutilate cattle and livestock, and modern UFOlogical accounts of cattle mutilation and disappearances, are obvious and do not have to be belabored. The resemblance between "kelpies" and modern descriptions of "lake monsters" and some "sea serpents," which are often described as having a "horse-like," "sheep-like," or "camel-like" head, is also apparent. As for actual evidence that cattle are abducted and taken to anomalous regions, there is one account in particular which supports this contention.

On Saturday, Sept. 2nd, 1989, the Associated Press posted a headline, which read "Marine Scientists Find A Sunken Cow." The dateline was Sitka, Alaska.

The gist of the article concerned a pair of scientists who were counting fish from a submarine some 20 miles west of Baranof Island, 690 feet beneath the surface of the sea. Tory O'Connell, a biologist with the state Department of Fish and Game, indicated that both she and the pilot of the two-seat submarine saw the carcass of a sunken, intact cow for a duration of several seconds as they cruised along the ocean bottom.

According to the AP article, the Sitka Sentinel sent a reporter to check out the video footage which the scientists had made of the scene, and he also reported that the object was not only a cow, but it was a Holstein, to be precise--a large, white breed of dairy cow with big, black spots.

Again, according the AP article, O'Connell can be heard on the tape counting off fish as the submarine creeps along the ocean floor at a depth of 690 feet.

The pilot, a man named Don Slater, could be heard on the tape stating:

> "What the hell is that?" Then, as the object comes into full view, "It's a cow." It is assumed that there was some degree of amazement and incredulity in his tone. According to O'Connell, the condition of the cow indicated that it "hadn't been there for long."

The AP article states that the pair offered no theories as to how the cow ended up on the ocean floor. Small wonder, as this enters the realm of the inexplicably improbable. The defensive wall of ignoring the data and downplaying the weirdness goes up, and the bizarre occurrence is quickly shelved by everyone, and all but forgotten.

Yet the question remains: How did an intact, "fresh" representative of a breed of dairy cattle, more at home in the temperate heartland of the U.S. and Canada, end up floating at THE BOTTOM OF A COLD OCEANIC TRENCH in the North Pacific? How did it get to its location, twenty miles from an island, hundreds or thousands of miles from home? Why was it at THE BOTTOM, instead of floating in a higher, warmer stratum of water? And how did it remain "intact" or in a fresh state, if it had somehow "washed out to sea" or floated from the North American continent? All interesting questions, especially when the "fairy" habit of stealing not just cattle, but milk and butter, is recalled. The Holstein is the ultimate milk-and-butter machine, after all. It should also be noted, strangely enough, that "Sitka" is a contraction of "Shee-Atika," meaning "People on the edge of Shee", an expression the original Tlingit Indians used to describe themselves by alluding to their location on the coast of what is today called Baranof Island--the significance, if any, being that the Gaelic Irish/Scottish underworld abductors of cattle, and thieves of milk and butter, were called the SIDHE, which is pronounced SHEE. Such strange coincidences seem to abound in the field of paranormal or Fortean research, which is in itself a very strange fact, hinting as it does of many underlying connections that we surface-dwellers only glimpse upon occasion. And of course, this entire episode with the "sunken cow" leads directly to a related subject: Unidentified Flying Objects, or as they're sometimes called, Unidentified Submarine Objects or USOs, which will be discussed further along.

In "East o' the Sun, West o' the Moon," the abducted maiden is rescued from a future of breeding with the denizens of the underworld, who shun the light of day.

Illustration: Donn P. Crane

2. WHERE DRAGONS ROAM

Human beings have always viewed the oceans with a mixture of fascination and dread. Early religious systems and mythologies populated the seas with hosts of monsters, demons, and mer-people. The Celtic people had their selkies, merrows, mermaids, and other humanoid sea-people, as did the Scandinavians; the Japanese had their "kappa," or aquatic, reptilian humanoids (which were said to subsist on "cucumbers and human blood"), and they and the Chinese both shared a belief in the Dragon-King of the sea and his daughter, sometimes called Otohime, who could appear as either human beings or as dragons. The Teutonic and Scandinavian peoples believed in undines, nixes and nixies, sea-trows, (sea-trolls,) and the dreaded "draug," or reanimated drowned man. Similar beings, such as the modern "Loveland Frog" of Ohio, have been seen all over the world and throughout history.

Alonzo resists the seduction of the Yara, a type of nixie or water-nymph, who has come up from a great depth. He resists by thinking of the woman he loves.

Illustration: H.J Ford

In mythic traditions (somebody else's archaic religious beliefs), the waters of the world are inhabited by nerieds, sirens, rusalka, vodnik, nicors, sea-dragons, Nagas, fairies, demons, jinn, and others too numerous to mention. Some are humanoid in appearance, ranging from hideously reptilian or deformed to incredibly beautiful and even seductive; others are depicted as great beasts of the sea, nicors and dragons that lurk under the surface, to snatch and devour. Some of the latter accounts are undoubtedly based on encounters with both known and unknown creatures of flesh and blood; others are symbolic of another truth, a twisted relationship between US and THEM, "them" being the nixies, Nagas, mermen, mermaids, and other beings rumored to have unhealthy appetites for the flesh and genetics of human beings and terrestrial livestock. Over time, the two types of encounters have merged into one archetypal mess of symbolism and mythic form; but the mess can be sorted out if looked at logically, and if all the evidence--even "traditional" or anecdotal evidence in the form of folklore and myth--is examined carefully.

Akkadian seal (circa 2300 B.C.) showing a group of lesser gods being brought before EA, Lord of All Waters. His throne is surrounded by the waters of the Earth.

The Biblical creation story itself starts with darkness over the primordial seas, or else a flooded world ready to begin anew; and the oldest human culture, that of Sumer, claimed that civilization, complete with agriculture, warfare, metallurgy, science, medicine, writing, and all the rest, came out of the seas. The Sumerians, who possessed knowledge which is in itself an anomaly, such as the existence and orbits of all the "spherical" planets around the sun (yes, all the planets, including Pluto and another, gigantic maverick world that modern science has yet to officially acknowledge), believed that the arts and sciences had been taught to them by the mysterious god EA, later called Enki, a giant who had scales like a fish or reptile and who dwelt in a traveling ship that could move beneath the sea as well as fly to the various lands--and heavenly abodes--of the Annunaki, or gods. Enki, interestingly enough, is the Sumerian equivalent of the

Hebrew version of the devil, responsible for the fall of the ADAM and his wife in the original "garden of the gods" which they tended. He orchestrated this fall by giving the ADAM and his mate the ability to reproduce sexually. He was depicted as humanoid, yet fish or serpent-like. The Phoenicians and Philistines worshipped this same deity, portrayed as having the upper body of a man and the hindparts of a fish or sea-serpent, as Dagon. To the Greeks he was the trident-wielding Poseidon, and to the Romans, Neptune. To the Sumerians as well as to later Mesopotamians, Ea/Enki was a Promethean figure, who had rebelled against the other "gods" and set humanity on the road of self-determination; but their view may be best represented as a dichotomy, for they also believed that EA was merciless at times, as were those who served him. One ancient prayer, as translated by R. C. Thompson, in his work THE DEVILS AND EVIL SPIRITS OF BABYLONIA, went thusly:

> *"'Seven are they, seven are they. In the Ocean Deep, seven are they. They are reared in the home of the Ocean Deep. Neither male nor female are they. They are as the roaming wild beast. No wife have they, no son do they beget. They know neither mercy nor pity. They harken not unto prayers and supplications. They are as the horses reared on the hills. The Evil Ones of EA, throne-bearers of the gods are they. They stand in the highway to befoul the path. Evil are they, evil are they. By Heaven be ye exorcised!'"*

Could this ancient passage mean that there are seven oceanic "kingdoms," or nations, in competition for the resources of the surface world? Is this related to the centuries- or millennia-old concept of the waters of the oceans being divided into the "seven seas?" Greek mythology seems to hint at this same secret, in the form of the Hydra, a seven-headed sea-dragon with a common body between them! The Hydra is dispatched by Hercules, and even lingered into fairytales and legends as "the seven-headed dragon" or the "seven-headed serpent--" which invariably came from the watery depths. Some variants of the hydra story give the beast nine heads instead of seven, but it's not difficult to see it as emblematic of an oceanic empire that is divided into sub-kingdoms. Hercules had other encounters with "reptilian" humanoids as well. He battled the shapeshifting ."river-god" Achelous, who had the foreparts of a man but the hindparts of a giant water-serpent, for the hand of the virgin (a code-word for genetically-desirable) Deianeira whom the "god" (for whom the river is named) fancied for himself.

The Seven-headed Serpent

A fairytale version of the Hydra.

Illustration: H.J. Ford

Could the genderless nature of these seven "beings" be behind their seemingly constant interest in and desperation for genetic materials, abduction of children, youths, and maidens, and the legends of seduction or rape? Or as reported more recently, genetic or medical tests involving sperm and ova, conducted by "gray aliens" and other types of unknown entities, who more often than not are described as genderless, drone-like, or androgynous? And of course there's "standing in the highway to befoul the path." How many modern UFO encounters and alleged abductions have taken place on dark, deserted highways, in the dead of night?

Another "god from the sea" is described by the Dogon tribe of Mali (a people possibly of Egyptian descent), who claim that their ancestors received instruction in the arts of civilization from the NOMMOS, a scaled humanoid species which came from the sea in submersible vehicle, and told the Dogon about traveling to other stars and planets. In fact, the NOMMOS pointed out to the Dogon the existence of a dark-star companion to Sirius-B, which wasn't discovered by western astronomers until 1862. These NOMMOS are essentially identical in form to the Sumerian culture-god OANNES and his fellows, who was either an aspect of EA, or was one of his emissaries. OANNES made his home in the Persian Gulf, while EA was Lord of All Waters, with a partiality toward large freshwater bodies (according to Berossus, or Bel-Usur in Akkadian, a Chaldean priest of Bel in Babylon who wrote circa 290 B.C.).

Often, UFOnauts and their henchmen-in-disguise, the MIB, are described by percipients as "wheezing," "gasping for breath," "staggering" as if drunk or unaccustomed to walking, often hairless and web-fingered, and with "bulging" or "frog-like" eyes. Are these symptoms of entities who are out of their accustomed greater pressurization, in the deep earth and sea caverns or ocean depths? Are they, in fact, "out of their element," "fish out of water" or "frogs on hot pavement," and suffering from the dread of all scuba-divers who depressurize too quickly, "the bends?"

Tannhauser and his underworld lover
(See Color Inserts)

Nearly everyone on Earth in this time of global civilization and global media is familiar with the enigma called UFOs. Vast numbers of these sightings and encounters occur over oceans, rivers, and large lakes, like the Great Lakes in the United States. Many reports exist of UFOs plunging into the depths of lakes and oceans, or bursting forth from the watery deep to zip away at alarming speed. Other encounters with gigantic, hovering craft, or dozens of smaller craft in formation, have been reported from seas and lakes around the world. The Navies of the United States, the Soviet Union, Brazil, Norway, the United Kingdom, and many others nations, have played hours-long or days-long games of cat-and-mouse with these objects as they would travel underwater at alarming speeds which are impossible for human submersibles or missiles. Battleships and submarines have pursued them in vain, and have even collided with them. Numerous cases of fishermen being abducted from docks and boats by UFOnauts have been reported even in recent years, and the high seas have a long tradition of strange lights, and passengers and crews vanishing from undamaged, eerily-empty ships. Such reports, in fact, go back for hundreds of years. These unknowns are generally termed USOs, or Unidentified Submarine Objects. Just a few of the hundreds of documented cases of these phenomena include:

1950s, the Black Sea: A UFO leaves the water and comes up behind a warship, witnessed by officers and crew. A photograph was allegedly taken.

August 1954: Captain Jan Bos, of the SS Groot Beer, enroute from Amsterdam to New York, reported observing a huge moon-like object rising out of the Atlantic ocean, about ninety miles east of Cape Cod light. It had illuminated portholes or windows around its rim, and departed at an amazing rate of speed.

In August 1965, a crew of the steamship Raduga, while cruising in the Red Sea, observed a brilliant, blazing sphere surfacing from under the water to hover over the surface of the sea, with great illumination. The sphere was estimated to be approximately one-hundred and eighty feet in diameter, hovering above the sea at an altitude of 450 feet. A gigantic pillar of water was created as the sphere emerged, to collapse afterward.

On October 4, 1967, shortly after 11:00 PM, a UFO around 60 feet in diameter was seen to over the sea near Shag Harbour, Nova Scotia. The UFO, displaying four bright lights which flashed sequentially, tilted at 45 degrees and plunged into the sea with a great flash and sound. Worried witnesses called the Barrington Passage RCMP, but most believed that a large aircraft had crashed into the harbour and were concerned about rescue operations.

Leviathan laid low affecting earth and sea.
Illustration by: Gustave Dore

Arriving near the scene, the investigating officers discovered that the UFO was floating on the water only a half-mile or so from shore, glowing with a yellowish light and leaving a trail of yellow foam as it drifted along in the tide. A Coast Guard cutter detached to the area arrived after it had submerged, but the foam was still present. Decades later, in 1993, MUFON investigator Chris Styles looked into the case, contacting several of the original witnesses and interviewing them. MUFON investigator Doug Ledger joined the investigation, and together they discovered that the incident had involved the both Canadian and U.S. Naval forces, over a period of several days.

According to ex-military witnesses involved in the case, the UFO that hit the water at Shag Harbour had departed as a submersible (USO) after it submerged. Traveling in this mode to a location some twenty-five miles away, it approached a joint U.S. and Canadian submarine detection facility, where it was detected. A small group of naval vessels was dispatched, located the USO, and took up positions above it. Finally, as a salvage operation was about to begin, the first USO was joined by a second underwater UFO, which some theorized had arrived in order to assist the first one with repairs. The U.S. Navy decided to wait it out and maintain surveillance, and after a week, some of the naval vessels on duty had to depart in order to investigate a Russian submarine which had entered Canadian waters. With the departure of some of the ships, the pair of USOs began to flee underwater toward the Gulf of Maine, at length outrunning the last of the naval vessels. The Navy had the distinct pleasure (or pain) of watching them leave the ocean and disappear into the sky at alarming speed.

Accounts like this exist from antiquity, although particularly good eyewitness reports have been documented from the 1800s to the present time. In fact, such cases could fill a small to medium-sized book, if one were to dwell on them at length. Suffice it to say that the flying yet aquatic dragons of the Far East and the dragon-disks of the ancient middle-east are still around, busy as always on their mysterious missions that circumvent the meddling of humanity.

The descendants of Abram of Ur, later called Abraham, a Mesopotamian of priestly descent, held a different view of the same marine being or species, a version at odds with some others. The Habiru or Hebrew view of this same personality-culture was very different, which of course became the prevailing view in Christian belief. They called the oceanic manifestation of Satanic power Leviathan, from Hebrew

LIVYATAN, the primordial sea serpent who was also identified by them as being another form of the "subtil" serpent of the garden scene. His source is in older Mesopotamian religious belief, where he is identical to the sea monster deity of Ugaritic myth called Yamm and Lotan (Leviathan), who warred with the highest god El and lost his rulership of the Earth as a result, just as Ea lost the same from his brother Enlil and father Anu, and as Satan lost his former station by rebelling against Yahweh, and in Christian belief, against a heavenly Christ as well. Leviathan appears in Psalms 74:14 as a multi-headed sea dragon of immense size which is defeated by God, and given as food to the Hebrews in the wilderness--or fed to the righteous on the Day of Judgment, depending on the interpretation. In Isaiah 27:1, Leviathan, possibly a real species of aquatic animal, is also a serpentine symbol of a spiritual, and perhaps physical, enemy or enemies of God:

In that day the LORD with his sore and great and strong sword shall punish leviathan the piercing serpent, even leviathan that crooked serpent; and he shall slay the DRAGON THAT IS IN THE SEA.

Not only is leviathan in the sea, but he is a "crooked" (wavy) serpent, a "piercing" serpent, and in other biblical verses, a "fiery flying serpent." Insignia of dragons and winged serpents have been reported by dozens or hundreds of witnesses to UFO encounters, as being displayed on a variety of uniforms worn by a variety of beings, ranging from essentially human to utterly outrageous in form. Another form of insignia which has been reported is that of an ouroboros, or a snake-like dragon with its tail in its mouth--identical in nature to the "Midgard Serpent" Jormungand, who encircles the whole Earth, yet lies on the floor of the ocean. Insignia signify two things, in this instance: a highly-organized society or societies with a strict hierarchical structure, and a militaristic or mission-oriented culture. One can only wonder if there are seven distinct variations on this theme, in accord with the seven seas and "evil ones," but that data has not been gathered and classified as of yet, in UFOlogical circles.

Satan as leviathan, imprisoned
beneath both the seas and earth.
Illustration: Gustave Dore

Psalm 74:13-14 states:

> *Thou breakest the heads of the DRAGONS IN THE WATERS. Thou breakest the heads of leviathan in pieces, and gavest him to be meat to the peoples inhabiting the wilderness.*

Battle of the Sea Gods, by Albrecht Dürer, c 1494. Based on an earlier drawing by Andrea Montegna, c 1493.
Note that each of the warring leviathans carries the genetic human prize upon his scaly back

Are the "dragons in the waters" the same seven beings, or principalities-nations, referred to in the Babylonian prayer?

Revelation Chapter 5, verse 13, presents an aspect of creation--an intelligent, presumably non-human aspect, with free will--which will be brought under subjection and forced to acknowledge Jesus Christ/Yeshua as the Hebrew God incarnate, and as ruler.

Presumably this takes place after the slaying of leviathan mentioned in Isaiah 27. Some of these rebellious beings dwell within the Earth itself, and others are beneath the surface of the oceans of the world:

> *And every creature which is in heaven, and on the earth, and under the earth, and such as are in the sea, and all that are in them, heard I saying, Blessing, and honour, and glory, and power, be unto him that sitteth upon the throne, and unto the Lamb for ever and ever.*

History seems to show that the human governments of the world, and human affairs in general, are greatly influenced by this "beast" and the "empire" or race it personifies (or perhaps indwells), and the last book of the Christian New Testament makes it clear that the fullest influence will be exercised at the "time of the end," shortly before the return of Yeshua or Christ to planet Earth. At that time, microchip implant technology (the "mark" of this same "beast," which brands human beings as livestock or property in an inventoried, monetary system), holography and mass electronic communications (the "image" of the "beast"), along with the horrors of a world dictatorship, will all be brought to bear upon an unrepentant humanity. The multi-headed beast which arises from the sea (Revelation 13:1-4), symbolizing the Mediterranean influence of a revived Roman Empire or European superstate, is no less symbolic of an unhuman power behind the scenes, which manipulates a chosen human as permitted ruler of a one-world government system. This dragon-beast, as multi-headed as leviathan was in Psalms 74, rises out the sea where it has been waiting and acting with dire influence throughout history, much like the Norse World-Serpent Jormungand, who rises from the depths at the end of the Nordic world-cycle. This rise is symbolic of the rise of the Antichrist (perhaps a genetically-suitable, even painstakingly designed, host for a non-human intelligence) and his empire, but it is just as indicative of the dark powers at work pulling the strings, hidden from the surface world of humanity.

This leviathan is indeed a formidable opponent and manipulator of humanity. It is the ruler of wealth and pride, of secrets, and possessed of strange abilities which could be described as either magical or technological, depending on one's point of view:

> *Upon earth there is not his like, who is made without fear.*
>
> *He beholdeth all high things: he is a king over all the children of pride.*
>
> *-- Job, 41*

And:

> *Out of his mouth go burning lamps, and sparks of fire leap out.*
>
> *Out of his nostrils goeth smoke, as out of a seething pot or cauldron.*
>
> *His breath kindleth coals, and flame goeth out of his mouth.*
>
> *--Job 19-21*

Is this a beast or a machine? A warship or submarine, a sea-serpent, or the devil himself? A proud and wicked civilization, which influences the proud wickedness of man? Or is it perhaps all of these things, symbolic of an ancient power and race of beings, spiritual and physical, who are akin or identical to those dark powers the ancients believed to be within the earth itself?

In his work LEVIATHAN (1650), Thomas Hobbes envisioned a sovereign state of immense power looming on the horizon, to dominate the human world:

> *"Hitherto I have set forth the nature of man, whose pride and other passions have compelled him to submit himself to government; together with the great power of his governor, whom I compared to LEVIATHAN, taking that comparison out of the two last verses of the one-and-fortieth of Job; where God, having set forth the great power of Leviathan, calleth him king of the proud."*

With the current talk of globalism, of a one-world government and monetary system, and the forboding images of the Apocalypse/Revelation, one wonders if Hobbe's vision of a cruel world superstate is near at hand. It would seem that his choice of title for his work was perhaps more fitting than he realized, as evidence indicates that a hidden, manipulative menace, indeed circling the world like an invisible constrictor as did the Norse world-serpent of Viking myth, has always been with us whether we want to believe it or not.

In centuries past, traveling "morality plays" would pass through towns and countryside, the finale featuring the "Hell Mouth:" a devouring reptilian dragon or devil's head, into which actors were forced by others in "hairy humanoid" or demonic garb.

Illustration: Donn P. Crane

(See Color Inserts)

3. WATER, WATER, EVERYWHERE

We scuttle about the surface of our planet, less than termites or waterbugs in a backyard spa. We know more about the surface of the Moon and Mars, in some ways, than we do about the deep ocean-bottoms and trenches, the "bottomless" crater lakes and mountain lakes of our own world. We are constantly amazed, as when we discover giant tubeworms living at great depths, under great pressure and boiling temperatures, drawing sustenance from chemical stews which spew from volcanic rifts in the ocean floor; we chatter excitedly at the few lesser specimen of giant squid and other monsters of the deep that we have acquired, or perhaps only glimpsed. We learn more every day about the machinery of global climate, which is ocean-driven, the delicate and intricate ecosystem and food-chain of the planet, which is ocean-driven, and numerous other oceanic marvels too numerous to name.

The earth and sea dragons, preying on that which they see as resource. By an unknown medieval artist.

We know nothing. We are blissfully ignorant. When something of true significance, true import is discovered, we turn our heads and pretend that we didn't hear; we laugh, scoff, or forget about it conveniently.

In Japanese legend, Uraschimataro is taken as consort by the Queen of the Sea, synonymous with the "Dragon-King's Daughter" of other tales. Growing homesick, he at length escapes back to the surface world, to meet the fate of many a guest of "fairyland." Ageing suddenly, he becomes an old man and dies. This "time dilation" or altering flow of time/missing time is a common theme to both folktales and UFOlogical abduction accounts.

Illustration: H.J. Ford

Tubeworms and giant squid are all very nice, but there are much stranger fish to fry, furtive beings dwelling beneath the secretive surfaces of our lakes, seas, and oceans.

In 1987, a group of scientists from Peking University were setting up to film wildlife in deep pools--unplumbed, really--near Wuhnan, Hubei Province. As they were setting up their cameras and cables, three gigantic beasts, with gaping mouths six feet wide, emerged from the water and began to swim toward the stunned men, who were led by Professor Chen Mok Chun. The group of nine witnesses described these creatures as "giant toads," pale and ghastly, who examined the men with hunger; one of the things shot out a gigantic tongue, wrapped it around some camera tripods, and devoured them. The other two suddenly screamed horribly, and then all three submerged into the unexplored depths of the pool (Dr. Karl Shuker, FATE, Sept. 1990).

Japanese kappa on holiday, or frog princes of huge girth? But kappa are said to be dwarfish, and the frog-prince more interested in meeting girls than terrorizing Chinese biology professors. An unknown, "cryptid" life-form, as yet unclassified? Whatever the case, substantive witnesses were involved, and the incident was well-documented and reported.

The British Oceanographer, Sir Anthony Laughton, reported a strange finding to the Royal Institute of Oceanography, based on careful examination of deep-sea photographs of huge imprints, which were taken at a depth of 12,000 feet in a deep Atlantic trench (Peter Kolosimo, NOT OF THIS WORLD, 1971). Upon considering the data acquired by a deep-dangling camera, Sir Anthony cautiously observed that the marks, marching away into the darkness, "remind us of the footprints of a biped." A biped of immense size, strolling under a weight of water that would crush an elephant like a grape? And what degree of additional weight would be required to drive deep prints into an ocean bottom already under tremendous pressure?

How many mysteries have been glimpsed yet have gone unreported, due to fear of ridicule or reprisal? Other anomalous structures have been seen in the deepest ocean depths, such as the "antenna array" glimpsed in a deep trench in the South Pacific some years ago, by the remote camera of a major government institution, which was dragged past the obviously-artificial device that appeared to be jutting by design from the ocean floor. This sighting, which was never explained, was reported briefly in small-headline fashion, then all but forgotten--or ignored. Other evidence of both past civilizations covered by the seas, or present ones which secretly thrive beneath them, have been found around the world.

Out of the sea, in search of the child....

Around the world from the South Pacific and in the Caribbean is another such anomaly.

Located less than a mile from the shores of Bimini and in less than fifteen feet of water is a series of titanic, rectangular stones, laid out in two parallel rows. This unexplained formation, a roadway formed by giants or the technology of giants, shoots off and away into one of the deepest zones of the Atlantic. The huge remains of toppled columns are found in the area as well, which have been confused by some with more recent debris which came from the sinking of a ship or ships carrying barrels of cement in powder form, which subsequently hardened into stone shapes on the ocean floor when the wooden barrels rotted away. The road and original columns, however, are beyond the explanation of conventional science and history.

An even greater mystery (World Explorer Magazine, Vol. 2 Number 3, 1998, Fortean Times, 114, Sept. 1998) has been discovered in recent years off the coast of Okinawa, Japan, documented by hundreds of divers and thousands of photographs. Near the island of Yonaguni, divers have located the remains of a cyclopean civilization, covering over 311 square miles, in eight separate locations beginning in March 1995. Recently, in 1996, another diver discovered a gigantic, angular platform some forty feet below the surface, this time off the southwestern shore of Okinawa. Divers look like tiny shrimp, swimming over and around its vastness. It is undeniably and irrefutably of artificial origin, as are the other structures, which consist of everything from platforms and terraces of such a size as to stagger the mind, usually formed of one piece of solid stone, to gigantic slabs of stone so cunningly fitted together that it would leave the greatest engineering minds perplexed. Straight streets, staircases, archways, perfectly-cut notches, have all been identified, yet they seem to all be made to accommodate beings of titanic, truly gargantuan stature, and the angles are strange and seem not designed for human or humanoid use. The stone is, overall, of a GREENISH hue. This is a substantial, obvious site of non-human civilization on our planet, at some time in the past! Why has this not been the major news story around the world that it deserves to be? Is this part of the kingdom of Otohime, and her father, the Dragon King of the Sea?

It seems that the American horror writer, H.P. Lovecraft, mentioned a site almost identical to this one, only located much deeper in the Pacific, in his "Cthulhu Mythos" cycle of stories: R'lyeh, a sunken city of gigantic stones, where the somewhat-crustacean, somewhat-reptilian, somewhat-molluscoid demon-god of the Pacific sleeps and awaits a terrible awakening. Cthulhu "lies dead yet dreaming," but with his mind, he sends unhuman servants of horrible type out into the world of men to do his bidding, and to corrupt men to his cause, the eventual revival of an age of terror on Earth. As for the description of R'lyeh, Lovecraft describes it in "*The Call of Cthulhu*:"

> *"He talked of his dreams in a strangely poetic fashion; making me see with terrible vividness the damp Cyclopean city of slimy green stone--whose geometry, he said, was all wrong--"*

Coincidentally or not, this is an exact description of the ruins at Yonaguni. What did Lovecraft actually know, or thought that he knew? For "Cthulhu" seems to correspond to the Sumerian/Akkadian "Kutu/Nergal," a demon of the underworld sea called the Abzu, and in Lovecraft's fiction, this being was summoned with chants containing cries of "Ia! Ia!" which is of course identical to the name of EA, "Lord of Waters." Is Cthulhu/Kutulu one of the regents, under "EA," of one of the "seven seas?"

Another mystery is that of the "globsters," immense masses of protoplasm, gelatinous yet covered in hair or cilia, which have washed up on beaches or have been drawn up in trawler nets, all around the world. Many have been photographed, with witnesses standing near them, but almost all either disintegrated or were thrown back into the sea by superstitious or nauseated sailors and fishermen. One of these doughboys washed up on a Tasmanian beach in 1962, several tons in weight, covered in "hair," its dimensions roughly 21 by 18 feet, and with boneless appendages which defied classification. At least the naturalist on hand at the time, Bruce Mollison, thought it did just that.... even after it dissolved into protoplasmic goo before any substantive amount could be preserved for analysis. Was this THING, somehow, a giant Cthulhu-like biped, under the proper deep-sea pressure conditions? Maybe Lovecraft wasn't as paranoid as some literary critics have observed....

Traditionally, wells and deep pools can be entrances to the underworld, as well as the abodes of subterranean or submarine beings.

Illustration: H J. Ford

Perseus rescuing the "virgin" Andromeda, and her genetic potential, from the "dragon" or leviathan of the sea.

Illustration: Willy Pogány

(See Color Inserts)

Anomalies like these make UFOs and Sasquatch seem as mundane as traffic noise. The fact that all such anomalies may have an inter-related or common origin is not heartening, however.

Fairy stories and folktales are filled with imagery which sheds further light on these mysteries, and demonstrate that, rather than being expressions of some nebulous "collective unconscious," or puerile attempts to explain the unknown, such stories may be a "collective defensive mechanism" of the human mind, by which realities that are too staggering or horrifying are downplayed and turned into fanciful entertainments, or are allowed to be dismissed as "figments of the imagination." It's worth considering that this reaction may have been conditioned to come into play in our minds, over time, both for our own psychological protection, and, unfortunately, to the benefit of those who perpetrate parasitical atrocities against the human race. In our present day, this "mechanism" kicks into effect all the time, as in the ridicule and mass-culture "iconification" of UFOs, "aliens," bigfoot, and the vampiric "el chupacabras" (which is not only similar to the UTUKKU and "scorpion-men" of the Sumerian underworld, but is also nearly identical to the quilled or spined "mardkhora," or man-eaters, of ancient Persia, and the European variant called a "manticore"), as well as a myriad of other terrifying marvels which defy reason. There is documented evidence of a link between the UFO phenomena and poltergeist activity, and a link with "channeling" of supposed extraterrestrial entities (a phenomenon identical to nineteenth-century "spiritualist" or "trance medium" interaction with ghosts, demons, sylphs or fairies), as well as a connection to pseudo-religious visions or visitations, which have spawned many a new sect or religious movement in the last two centuries alone. Also worthy of note are the alleged telepathic abilities of "alien" beings, which in an earlier age would have been called the "language of the spirits" or the "language of angels." While the various strange manifestations are at times physical in nature, there is a distinct and observed "paranormal," spiritual, or other-dimensional (quantum) aspect to all aforementioned unexplained phenomena. The human mind is not always capable of dealing with unquantifiable activities such as these. Those things which frighten us are ridiculed and hence minimized in import or their ability to frighten us, to our minds. If they are simply too perplexing to be dealt with in this fashion, then they are mentioned or published briefly and are afterward ignored and "forced into the state of being forgotten," as has been pointed out already.

Fairytales are filled with mermaids and mermen, nixies, water-dragons, and dozens of other "fanciful" entities which engage in abduction, seduction, and cross-breeding of and with human beings. Put simply, sex and reproduction, followed closely by predation for purposes of satisfying hunger (particularly for blood and other fluids), are the prevalent theme in ALL such stories from all the cultures of our planet. Sea-dragons, and dragons in particular, seem to have an inordinate hunger for "virgin" flesh, as if ordinary flesh--that is, that which might already be carrying human seed to fruition--is not good enough to satisfy their "hunger," which is more often than not thinly-veiled sexual or reproductive appetite. Human children, human sexual partners, human midwives and wet-nurses--all are abducted, apparently, in accord with some ongoing program or agenda which spans centuries upon centuries.

Sometimes the hero must identify the abducted by picking from a line-up of duplicates, or clones. A similar scene takes place in the Mayan underworld of Xibalba, as recounted in the Popol Vuh.
Could abductions solely for cloning purposes have taken place?
Illustration: H.J. Ford

The present-day equivalent is obvious, in the form of reports of "alien abduction and genetic crossbreeding," "hybrid infants," and the repeated abductions of the human mothers--ova donors, really--for purposes of nursing, bonding, and nurturing ("alien" babies are pathetic, sickly, "wizened," and emaciated, just like the fairy-tale "changelings," fairy infants, or fairy-human crossbreeds).

Fairytales are also replete with accounts of "identical replacements" for people, called changelings, who, upon being discovered or somehow discomfited, wither into a sickly imitation of the person impersonated. Similarly, "doppelgangers" or doubles, sometimes called "the fetch," are numerous in Northern European accounts. One of the recurring themes is the confusion of a would-be rescuer who is determined to save someone from abduction to the underworld of Faerie; the rescuer is confronted with a group or line-up of beings identical in appearance to the abductee, and he must choose the genuine human being from the line-up. All of this begs an obvious question about the nature to which purloined genetics might be being put, if such abductions have indeed been occurring over the centuries, and it brings into question whether the process of "cloning" is as recent as is currently believed. What would be the ultimate purpose of such cloning and crossbreeding experiments, if not to isolate and perpetuate certain genetic characteristics or abilities?

In Hebrew esoteric tradition, the oceanic equivalent of the demoness Lilith is Naamah, a seductress of men and devourer of children. Either she's been really busy for three thousand years or so, or she's got cousins and descendants around the world.

Folklore teems with images of seduction of humans or "mortals" by beings from the aquatic realm. Some undersea folk were apparently more human-like than others. According to Irish folklore, a woman of the sea would sometimes shed her aquatic form and become the wife of a man, if he could steal or otherwise hide her "cap." As long as she was without this headgear, she could not return to the sea; but even if happily married for years, if the headgear were ever found, the murduchu, or mermaid, would make haste to rejoin the undersea kingdom she had come from. The implication is obvious: Without her headgear and breathing apparatus, she could not return beneath the waves, and the "shedding" of her aquatic form could have been the removal of some sort of diving suit.

Similarly, in Ireland and Scotland, the were-seals called selkies (sealchies) could only be captured by mortal men if they could manage to steal their "seal-skins" and hide them. A selkie wife behaved in a manner identical to a murduchu one. Also in Ireland, a common form of "water-spirit" was the bean-nighe, a relative of the bean-sidhe (banshee). The bean-nighe could supposedly be seen beating bloody rags on stones in the rivers and creeks that they haunted, and were believed to dwell beneath the ground.

The sea-dweller reclaims her "magic" technology, in this case a spinning-wheel.

Illustration: H.J. Ford

In ancient and medieval Wales, the Gwragedd Annwn were beautiful temptresses who arose from enchanted lakes to take their pick of mortal husbands, who would often be led beneath the lakes, never to return. This demonstrates the link between all underworld beings, whether of the land or sea, as Annwn (pronounced "ah-noon") was the Welsh underworld, filled with all variety of human-like and unhuman forms.

THE MERMAID ASKS FOR THE KING'S CHILD

The original caption says it all.

Illustration: H J. Ford

Amazing how the "wee folk" usually choose to appear to those who are young and potentially fertile.

Illustration: H.J. Ford

True to underworld form, if not after sexual intercourse with human beings, other beings wanted children, allegedly as part of their diet, but perhaps to also provide a captive genetic stock for their own isolated underworld settlements. In Britain, two nasty ladies of this description were "Peg Powler" and "Jenny Green-teeth," who, while female, were nevertheless of distinctly reptilian description. These monsters would grab children, particularly when isolated or alone, and drag them down into their underwater, underworld domains.

Some medieval accounts have merfolk creeping ashore in human guise at night, to seduce human beings, or to join in local festivities as a friendly stranger. One of the great tragedies of medieval folklore was that of Melusine.

Many variants of the Melusine legend exist, but the plot goes something like this: A nobleman (or youth, or king--an eager male) is taken with a beautiful maiden of mysterious origin. She agrees to marry him, but only on the condition that she have one day a week to depart and be alone. He agrees, and all goes well for a time, with children (half-human hybrids) and a generally happy home. Alas, one day curiosity wins out, and he follows surreptitiously as she sets out alone. Then he sees her either turn into a mermaid, a split-tailed mermaid, a dragon, or a toad-headed dragon-like monstrosity. Regardless of which it is, the honeymoon is pretty much over, and when he confronts her or slips up and reveals his knowledge, she departs in either a rage, or in great sadness. Some variants have her departing into a mountain rather than into the sea or river, and to some she reveals the enormous wealth she has accumulated there, over the centuries. During the middle ages Melusine or Melusina was believed to have been a real person (although not a human one), and some people attempted to claim lineage from or kinship with her, through her hybrid children!

In the story of Undine, written by the Frenchman La Motte Fouque, and published in German in 1811 (translated into English in 1818), a story almost identical to that of Melusine is recounted. But in this variant, the nymph or water-spirit Undine, by virtue of holy matrimony, gains a human soul and salvation, something her race lacks. This desire for salvation is a common theme in tales of water-dwelling beings. As a result of revealing her true origin to her husband, the knight Huldbrand, she loses him to the arms of another, a maiden named Bertalda (not so easy being human after all, eh?). Undine returns to her aquatic home, but returns on the marriage-day of Huldbrand and Bertalda, killing the knight with a venomous kiss.

Moving into the misty realm of pseudo-history, the Merovingian kings of France traced their lineage to Merovee, the founder of the line. It was believed that he was of human descent--on his mother's side this time--but while carrying him in her womb, she had been raped or seduced by a sea-being called a "Quinotaur," an entity which sounds very Oannes or EA-like, indeed. Although raised by human parents (his human father was Chlodio, the first king of the Franks) there was apparently something about Merovee which made them suspect he was not wholly human, and his name bears this out, for one meaning of it is "son of the sea." Another meaning, however, is "son of Mary," and the heresy of the Merovingian line, which set it directly against the Catholic Church, was that somehow they were descended, through Merovee, from a secret union between Jesus Christ and Mary Magdalene! But even with this somewhat historically and scripturally unsupportable claim, the Merovingians still also held to the belief that they were descended from a horny sea-god, the "Quinotaur." Actually the whole thing was pretty fishy, and the reason for the offspring of the "Quinotaur" to make such spurious claims, and to create centuries of divisions and suspicions between nations, creeds, and the medieval Church, will perhaps become obvious further along in this endeavor.

Ah, so much for fairytales, one might say. The problem is, people are still seeing these things.

On June 15, 1608, while sailing north of Russia near the Novaya Zemlya islands, famed navigator Henry Hudson inscribed the following sighting in his logbook, as reported by amazed members of his crew on watch:

> *"From the navel upward, her back and breasts were like a woman's... her body as big as one of us; her skin very white; and long hair hanging down behind, of color black; in her going down they saw her tail, which was like the tail of a porpoise, and speckled like a mackerel."*

A medieval view of an ocean-dweller.

So, obscure drunken sailors and fishermen were not the only ones to see strange humanoids in the ocean. On January 4, 1493, during his first journey to the Americas, Christopher Columbus made an entry in his journal that he himself, along with his crew, had witnessed a trio of mermaids breaking high from the sea off the coast of Haiti, going so far as to say:

> *"They were not as comely as they are painted, but to some extent they have a human appearance about the face...."*

In one African tale, "The Woman and the Ogre," the water/cavern-dweller follows the typical pattern for his type, in this case demanding the woman's child.

Illustration: H.J. Ford

Columbus also swore that he had seen the same types of beings on a previous voyage off the coast of Guinea, West Africa. It is doubtful that he saw stitched-together fish-and-monkey corpses thrashing about in the sea.

The 17-century explorer of America, John Smith, claimed to have seen a mermaid in an up-close encounter, on a trip through the West Indies in 1614. He initially saw her swimming and thought her a lovely maiden, but when he prepared to talk to her from the ship, he was shocked to discover that, from the waist down, "the woman gave way to the fish."

Obviously, she was not a fish. But what are these beings? Variants of the human family, returned to the sea in the manner of cetaceans? Human pinnipeds? Or, like the Irish murduchu and selkie, are they possibly wearing some sort of swimming apparatus, some manner of form-fitting technology, which gives them the appearance of a fish or porpoise from the waist down? The Sumerian and Babylonian ocean-gods of elder days could "shed" their aquatic forms and appear more human, as well. Hundreds of other eyewitness accounts have been preserved, from churchmen, sailors, housewives, medical doctors, others, in both the Eastern and Western Hemispheres. A most comprehensive and balanced report on the actual sightings, in more or less modern times, of the mermaid phenomena can be found in the book Unexplained!, by Jerome Clark.

So, if we go by all accounts--mythical, folklore, anecdotal--what can be determined to be the primary interest that these "merfolk" have in us? The answer should be obvious, and identical to the "needs" of other mysterious beings: intercourse, offspring, marriage, children (hybrid or abducted), and even the spectre of human flesh as food or other resource....in short, Genetics.

There also seems to be an inter-relationship between the "fairies" or "gods" of the sea and land, which is sometimes a rivalry. The cause of contention or cooperation between these two groups boils down to one resource: humanity itself. These "unseen empires" work together toward a more or less obvious common goal, involving use of their own and human genetics in efforts to obtain a result which may take centuries: the integration of their own "seed" or genetic offspring into the sunlit surface world, invisible among humanity by virtue of its seemingly-human appearance.

A medieval view of a merfolk couple. He leers, she pulls away. Who can blame her for going on shore-leave, if that's how they grow 'em at home?

Universally, folklore demonstrates two distinct groups of "dragons," those of the land, and those of the sea. Interestingly, the Bible does the same, describing a "dragon of the land" with almost as much descriptive prose as it does leviathan, and in Job 40:15-24, we see:

"Behold now BEHEMOTH, which I made with thee; he eateth grass as an ox. Lo now, his strength is in his loins, and his force is in the navel of his belly. He moveth his tail like a cedar: The sinews of his stones are wrapped together. His bones are as strong pieces of brass; his bones are like bars of iron. He is the chief of the ways (largest of the creatures) of God: He that made him can make his sword to approach unto him (i.e., only the might of

God can destroy him). Surely the mountains bring him forth food, where all the beasts of the field play. He lieth under the shady trees, in the covert of the reed, and fens. The shady trees cover him with their shadow; the willows of the brook compass him about. Behold, he drinketh up a river, and hasteth not: He taketh it with his eyes: His nose pierceth through snares."

The 'water-dwelling' subterraneans-nagas, trolls, fairies, or mermen – reward the girl for her 'attentions,' which are represented as 'grooming' and a kiss. Some girls will do anything to get ahead. Illustration by: H.J. Ford

Without a doubt, this describes no historically-documented creature, but it does sound suspiciously like an extinct variety: a sauropod dinosaur, like an apatosaurus, brachiosaurus, and so on. It also sounds like many a Norse and Teutonic description of a dragon, which more often than not, DWELT WITHIN THE EARTH. As in the case of leviathan, behemoth apparently symbolizes a living creature of huge size, an evil man, and an EARTHLY EMPIRE or religion that would someday come into partnership with leviathan to rule over mankind (Revelation 13:11):

> *And I beheld another beast coming up OUT OF THE EARTH; and he had two horns like a lamb (false or deceptive Christianity), and he spake AS A DRAGON.*

According to ancient tradition, as the Creator made all beasts male and female, he did likewise with Behemoth and Leviathan, the monsters of the ocean and of the earth respectively, making one male and one female of each (could one society be patriarchal, the other matriarchal?). But the mating of the earth dragon-beast, and the oceanic dragon-beast, was forbidden by God, lest they destroy the world. So much for legends, for the two "nations" or beasts have been working together to some extent for some time, and will supposedly do so to a greater degree in the near future.

A hint of this partnership can be found in medieval images of a type called "La Femme Aux Serpents," which are manuscripts or church and cathedral carvings depicting a woman suckling dragons or serpents, or being bitten upon the breasts by them, with her own feet sometimes turning into such creatures, or into the split-tail of Melusina or another mermaid/merdragon-type female.

This type of figure symbolized church corruption, lewdness, lasciviousness, and other sins of a lustful and promiscuous nature, but it had a deeper meaning. It also symbolized the underlying reality of medieval history--of ALL history---that humanity is ceaselessly beset by the parasitical attentions of the genetic-raping fertility-lusting, life-draining kingdoms of the twin dragons of Earth and Sea. And it symbolizes the tragic, seemingly-doomed, and perhaps corrupted nature of their victim, humanity.

Ah, youth! Seduced by a deep pool, by a non-human enchantress. The kids might not be what he expects, though.

Illustration: H.J. Ford

When leviathan and behemoth combine their efforts, the result is the reign of the Antichrist, who is controlled by an evil intellect of non-human origin, that same which rules the unseen "principalities and powers." Seemingly a "man" in form and nature, he is destined or designed to be "possessed" by Satanic power in order to revive the kingdom of Satan/Ea/the Old Gods on Earth. Is the creation of this "person," identified as not truly human by his appellation "the beast," and others like him, the eventual goal of all the "fairy-alien" crossbreeding experimentation? At some point after his rise to power, he declares himself to be a "god" before all mankind. This is demonstrated even now by the same "spirit" or attitude of "antichrist" which leads men and women to say "we are our own gods," "we're all part of god/the universal mind," and so on, sentiments which are also (coincidentally?) the same expressed by many a so-called "alien" during UFO encounters, when asked by abductees about their religious belief-systems.

Again, Lovecraft may have hinted at this in "The Call of Cthulhu," where the consequences of human beings seeing themselves as being "as gods--" even false gods--are predicted:

> *"That cult would never die until the stars came right again and the secret priests would take Cthulhu from His tomb to revive His subjects and resume His rule of earth. The time would be easy to know, for then mankind would have become as the Great Old Ones; free and wild, and beyond good and evil, with laws and morals thrown aside and all men shouting and killing and reveling in joy. Then the liberated Old Ones would teach them new ways to shout and kill and revel and enjoy themselves, and all earth would flame with a holocaust of ecstasy and freedom."*

The Earth might yet flame, indeed, as in Revelation 8:7-11, when God removes all of his restraint from mankind:

> *7 The first angel sounded, and there followed hail and fire mingled with blood, and they were cast upon the earth: and the third part of trees was burnt up, and all green grass was burnt up.*
>
> *8 And the second angel sounded, and as it were a great mountain burning with fire was cast into the sea: and the third part of the sea became blood:*
>
> *9 And the third part of the creatures which were in the sea, and had life, died; and the third part of the ships were destroyed.*

10 And the third angel sounded, and there fell a great star from heaven, burning as it were a lamp, and it fell upon the third part of the rivers, and upon the fountains of waters;

11 And the name of the star is called WORMWOOD; and the third part of the waters became wormwood; and many men died of the waters, because they were made bitter.

The devastating fire, translated into English as WORMWOOD, falls with flame and poison upon the world. But this may have already happened, in much smaller, preliminary form, for the Russian word for wormwood is....CHERNOBYL, or nuclear radiation and devastation.

Or, as Jesus Christ described what was to come in the Gospel of Matthew, Chapter 24, verses 21-22:

21 For then shall be great tribulation, such as was not since the beginning of the world to this time, no, nor ever shall be (again).

22 And except those days be shortened, there should no flesh be saved: but for the elect's sake those days shall be shortened.

In other words, the near-total nuclear destruction of mankind.

4. THE PERPETUAL REBELLION

So what's it all about? How do UFOs, USOs, cryptid and reptilian monstrosities or bipedal saurians, aliens and fairies, lost, sunken, and prehuman civilizations, titan-like suboceanic creatures, megalithic ruins on Earth, on Mars, and perhaps elsewhere, all fit together into one unifying theory, pattern, or puzzle-picture? Why do these things exist or seem to exist, and why do they seem to vastly predate humanity on this planet or in this solar system, and to defy all human knowledge and reason, most of the time? Why do mysterious beings prey upon humanity, seeking to interbreed with us, terrify us, mislead us, and eradicate us?

Answers to this exist in many ancient texts, but are aptly described and referred to in the Torah and New Testament of the Bible.

In the original Hebrew texts, the word YOM is used for the word "day" in the Genesis creation story. Sources as widely-disparate as the Sumerian scholar Zecharia Sitchin and Christian evangelist scholars who read ancient middle eastern languages, are generally in agreement that this word in its original context actually means "epochs of time," "ages," or something similar. Also, lost continents and civilizations are distinctly mentioned, as is their destruction. Many biblical references describe an earlier age, and the fate of pre-human civilizations which have vanished from the Earth. Some of these were overthrown by cataclysms which thrust them to the bottom of the ocean, or else were driven or plunged deep beneath the crust of the world. Such places are referred to as a "terror," and came about due to great planetary upheavals. References are made to the prehumans, the "old people" or "people of old time."

Ezekiel 26:19-21 states:

19 For thus saith the Lord GOD: When I shall make thee a desolate city, like the cities that are not inhabited; when I shall bring up the DEEP upon thee, and GREAT WATERS SHALL COVER THEE;

20 When I shall bring thee down with them that descend into the pit, with the PEOPLE OF OLD TIME, and shall set thee in THE LOW PARTS OF THE EARTH, in places desolate of old, with them that go down to the pit, that thou be not inhabited; and I shall set glory in the land of the living;

21 I will MAKE THEE A TERROR, and thou shalt be no more; though thou be sought for, yet shalt thou never be found again, saith the lord GOD.

It would seem that grays, reptilians, UFOnauts of many descriptions, and many cryptid critters which defy rational explanation such as "hairy humanoids," translated as "satyrs" in the King James version of the Torah, and "el chupacabras," fit the description of being "a terror," at least according to many who have had encounters with them. The statement "thou shall be no more" refers to a surface-world, pre-cataclysmic kingdom. There's much additional Biblical information to support this theory, as in the book of Jeremiah.

Jeremiah Chapter 5 contains a reference to the pre-human world that existed before the Creation as described in Genesis. Verse 23 begins with a description of the aftermath of destruction, where Genesis picks up; then the following verses describe a land which existed before, and which had both cities and living creatures (an ecosystem), yet no human or sentient beings ("man"), due to some cataclysmic event. A "time of darkness," akin to nuclear winter or the similar state which would have followed the dinosaur extinction event, is mentioned as well. Also mentioned is the fact that this was not "a full end," but life and civilization were destined to someday emerge again:

> *23 I beheld the earth, and, lo, it was without form, and void; and the heavens, and they had no light.*
>
> *24 I beheld the mountains, and, lo, they trembled, and all the hills moved lightly.*
>
> *25 I beheld, and, lo, THERE WAS NO MAN, and all the birds of heaven were fled.*
>
> *26 I beheld, and, lo, the fruitful place was a wilderness, and all the cities thereof were broken down at the presence of the LORD, and by his fierce anger.*
>
> *27 For thus hath the LORD said, The whole land shall be desolate; yet will I not make a full end.*
>
> *28 For this shall the earth mourn, and the heavens above be black: because I have spoken it, I have purposed it, and will not repent, neither will I turn back.*

In esoteric Hebrew traditions, Satan/Lucifer was the ruler of the Earth and surrounding planets before his rebellion and Fall, having been given the office of caretaker and supervisor over activities here in this world. Ambitious pride and jealousy entered his heart, however, and he launched a coup d 'etat, which failed miserably. A spiritual conflict, undoubtedly; a galactic war?...most likely. Ezekiel 28:14-17 describes the fall of Satan, and his interplanetary empire ("the stones of fire" are traditionally the rocky planets within the solar system) before the War in Heaven, complete with commerce (merchandise) and mismanagement or rebellion in the form of war or strife:

14 Thou art the anointed cherub that covereth; and I have set thee so: thou wast upon the holy mountain of God; thou hast walked up and down in the midst of the stones of fire.

15 Thou wast perfect in thy ways from the day that thou wast created, till iniquity was found in thee.

16 By the multitude of they merchandise they have filled the midst of thee with violence, and thou hast sinned: therefore I will cast thee as profane out of the mountain of God: and I will destroy thee, O covering cherub, from the midst of the stones of fire.

17 Thine heart was lifted up because of thy beauty, thou hast corrupted thy wisdom by reason of thy brightness: I will cast thee to the ground, I will lay thee before kings, that they may behold thee.

According to Judeo-Christian tradition, there was an original war which resulted in the expulsion of Satan, and one-third of the heavenly host (army), from the "ultimate" ("third") heaven or region of existence where God Himself resides and reigns (for the sake of the study at hand, it is not appropriate to attempt to define or elaborate on the nature or definition of God, other than to say that He is and was the ruler of this continuum). After this ill-considered attempt, Satan was curtailed in his activities and travels more or less freely within the region of the Earth and its surrounding region of space (the solar system and some space beyond, called "the first heaven"), for the duration of an expansive, yet ultimately limited, span of time. At some point even this will be taken from him, however, and he and his "heavenly" or galactic army will be driven earthward to join with his supporters and captains--those Nephilim or "fallen ones" who are already literally "in the earth--" to prepare for the final battle, their collective last stand:

And the great dragon was cast out, that old serpent, called the Devil, and Satan, which deceiveth the whole world: he was cast out into the earth, and his angels were cast out with him (Rev. 12:9).

The increase in UFO sightings and encounters over the last century is surely significant. Retreating armies in defeat (the "spirit wickedness in high places" or the heavens, of Ephesians 6:12), driven back to a final bastion where other defeated partners have long-dwelt in reserve, and have held invisible influence over human civilization (the "hidden kingdoms" of aliens/fairies, unhuman beings and their ilk, the "principalities and powers" and "rulers of the darkness of this world," also from Ephesians 6:12), are at the bitter end of millions of years of a galactic and spiritual war, and have been arriving in droves for decades, with a long-planned deception in mind. Revelation 12:12 states:

> *Therefore, rejoice, ye heavens, and ye that dwell in them (the victorious armies or hosts). Woe to the inhabitants of the earth and the sea! for the devil is come down unto you, with great wrath, because he knoweth that he hath but a short time.*

With an intergalactic and perhaps interdimensional (spiritual) war of millions of years of duration nearing its grand finale, ancient and sacred texts around the world make it clear that the Victor of this staggering conflict will follow the losers to their final desperate refuge, defeat them, and occupy the final planet in order to insure an everlasting victory.

References to things and events in various traditions and scriptures which could be considered spiritual, technological, or both, indicate that this final showdown will involve wonders such as human beings have never seen. The stage is indeed set for what has long been termed the "great end-time (current epoch of planetary history) deception."

What form will this deception take, in all likelihood?

Modern UFO abductees, contactees, "channelers," "remote viewers," and so forth, claim to have been contacted or informed by a mysterious group of "ascended masters" or "evolved" extraterrestrials, called "Ashtar Command," "Galactic Council," and other such lofty names or designations. Often such "entities" enlighten their adherents with the message that they are "special," "chosen," or use similar types of psychological and ego-manipulation. These people are often told that they are descended from the "sky gods," or are of hybrid ancestry, and are told that they have both a "dragon" or "reptilian" heritage, as well as a "Solar Christ Consciousness" the latter term being contradictory to every tenet of the teachings of Christ himself. In fact, many of those who claim this elevated "Christ Consciousness," which essentially means "I have a Christ Nature in and of myself," express dislike or even hatred for mainstream Christians and Christianity. The "ascended masters" have gone even further, for divisions are being drawn by these beings between themselves as the descendants of the Sumerian god ENKI (who is EA), and the followers of ENLIL, ENKI's older brother and "rightful heir" to rulership of the Earth. It's as if a propaganda machine is in full force, well in advance of some anticipated event.

The name "Ashtar" should be indicative of the true identity of the personalities behind these contacts, as this is a variant of the name of the goddess Ishtar, a rebellious enemy of heavenly authority in Sumerian tradition, and it is the same as the biblical Ashteroth, a demonic power or being which enslaved the will of its worshippers with sexual and sensual excess. To add to the absurdity of this ludicrous charade of misinformation, the "command" or "council" supposedly consists of not only non-human "aliens," but also such personages as Melchizedek, Mohammed, Gautama Buddha, and "Sananada," who also identifies himself as the same person as Christ. All of this is given some sort of "validity" in the mind of the human percipients by the strange events or forces which seem to accompany such communications. The similarity of "Sananda's" name to Satan does not, however, go unnoticed, nor does the admonition from Jesus himself, in Matthew 24:5:

> *For many shall come in my name, saying, I am Christ; and shall deceive many.*

And, in regard to the seemingly amazing origin or transmission of these communications from "Sananda" and his partners in deception, as well as any manifestations or appearances of their (either very real or holographic/illusory) vehicles or forms, in Matthew 24:24, we read:

> *For there shall arise false Christs, and false prophets, and shall shew great signs and wonders; insomuch that, if it were possible, they shall deceive the very elect.*

One of the constant themes which seems to emerge from various human adherents to these New Age "teachings," who nevertheless believe themselves to be "in the know" on the basis of their experiences (whether or not the experiences are truthful or are elaborate deceits doesn't seem to matter much), is that a superhuman galactic race called the "Els," "AngELs," "Atlans and Titans (according to Richard Shaver)," "Lyrans," "Vegans," and/or reptilian humanoids (called both "Galactic Man" and the "Serpent" or "Sky-People" by Brinsley Le Poer Trench; and who lately claim to have originated on Earth, justifying their return to reclaim their homeworld, OUR planet) are all returning to this world, for a variety of reasons. This may in fact be the case, but their reasons for returning supposedly include, but aren't limited to, the following, or any combination thereof:

1. To save humanity or "chemical man" (Le Poer Trench's term) from its own self-destructive nature, as we are a behaviorally-aberrant and violent species which they, the returning "sky-gods," claim to have created artificially or otherwise guided along the evolutionary path to sentience and civilization;
2. To save the ecosystem of the "Earth and Sea" from humanity, and thus save humanity indirectly;
3. To usher in a "New Age" (of Aquarius, another symbol of EA or Satan) of one-world governmental and societal (false) harmony and peace, under THEIR direction, leadership, and supervision (just as Lucifer ruled the primordial Earth before his first defeat, and just as the "human" leader and underworld stooge called Antichrist--his earthy form prepared by the hidden rulers of Earth to provide habitation for a returning puppet-master of greater, non-human intellect and spiritual power--is predicted to do, in the Books of Daniel and Revelation);
4. To help humanity through a coming "ascension" to a higher state or "vibrational frequency," which could be used pretty much as an excuse for anything, including genocide and execution; this is referred to in New Age and UFO channeling and contactee circles as "entering fifth density," and leaving our "inferior" three-dimensional reality; and it's also called "entering the photon belt," a region of space through which our solar system will supposedly pass (according to the same "ascended masters"), which in terms of actual scientific meaning or effect, is meaningless.

5. To help prepare humanity, the long-lost or long-nurtured (take your pick) offspring or experiment of the "aliens," for ANOTHER invasion, from a supposedly hostile force "out there" in the galaxy which is on its way to Earth. This latter point is especially pertinent, as it has direct bearing on the actual purpose of all elements of this Grand Deception:

The return of the Sovereign of this and all worlds will be met with a savage resistance, for humanity, for the most part, is to be pressed or deceived into the service of the world dictator and his New Age advisors (human ones, and possibly his newly-arrived, more "highly-evolved" "alien" friends, or their human mouthpieces), detailed in Revelation 19:19 and 19:20:

> *19 And I saw the beast (Antichrist, the leviathan/behemoth-produced puppet), and the kings of the earth, and their armies, gathered together to make war against him that sat on the horse, and against his army.*
>
> *20 And the beast was taken, and with him the false prophet (New Age/UFO cult religious leader) that wrought miracles before him, with which he deceived them that had received the mark of the beast (microchip-implant identity and currency voucher of a global society), and them that worshipped his image. These both were cast alive into a lake of fire burning with brimstone.*

Who is "he upon the white horse," and from whence does he come? The "white horse" is perhaps a brilliant aerial vehicle, as Revelation 19:11 tells us:

> *11 And I saw heaven (the skies) opened, and behold a white horse; and he that sat upon him was called Faithful and True, and in righteousness he doth judge and make war.*
>
> *12 His eyes were as a flame of fire, and on his head were many crowns,; and he had a name written, that no man knew, but he himself.*
>
> *13 And he was clothed with a vesture dipped in blood: and his name is called the Word of God.*

14 AND THE ARMIES WHICH WERE IN HEAVEN followed him upon white horses, clothed in fine linen, white and clean.

15 And out of his mouth goeth a sharp sword (a command of destruction?), that with it he should smite the nations: and he shall rule them with a rod of iron: and he treadeth the winepress of the fierceness and wrath of Almighty God.

Prior to this final and decisive conflict, human war (due at least in part to unhuman or spiritual manipulation) will be ongoing for approximately three and one-half years in the region of the middle east, complete with what can only be atomic and biological weapons (Daniel 9:27, 11:11, 11:44, 7:24, Ezekiel 38:15-16, Revelation 13:1). The bloody destruction of this unrelenting war will be unparalleled and will involve a marshalling of human national armies against one another, and then against a "new" invading army from the skies (Psalm 2:2, Isaiah 34:2, Zechariah 14:3, Revelation 9:16). This is ended by a LITERAL and PHYSICAL LANDING of a new, seemingly extraterrestrial invasive force on the Mount of Olives (Zechariah 14:4, Revelation 16:14), the same "force" that current "extraterrestrial channelers" and their unseen masters want to "warn" us about, and from whom they wish to "protect" us.

According to Scripture, all of this will be proceeded by the mass sightings of unknown objects in the skies, as recently occurred with UFO flaps in Mexico City and elsewhere; and perhaps by the return of an astronomical object that few on Earth yet know about, a "dark planet" that inhabits the darkness beyond the edges of our solar system, the gravitational effect of which may create upheaval and terror on Earth.

Jesus, in Luke 21:11 states:

And great earthquakes shall be in divers places, and famines, and pestilences; and fearful sights and great signs shall there be from heaven.

Luke 21:25-26 elaborates:

And there shall be signs in the sun, and in the moon, and in the stars; and upon the earth distress of nations, with perplexity: the sea and the waves roaring;

Men's hearts failing them for fear, and for looking after those things which are coming on the earth; for THE POWERS OF HEAVEN SHALL BE SHAKEN.

Tracking perturbations in the orbits of the outer planets in 1983, JPL and NASA determined that there was, indeed, a large, unknown body with an unusual orbit somewhere out beyond Uranus and Pluto. Earlier, in 1982 NASA themselves actually admitted the possible existence of an object they called "Planet X," orbiting beyond the outermost planets of our system.

In 1983 a newly launched system called IRAS (Infrared Astronomical Satellite) quickly found Planet X. The Washington Post quoted the lead IRAS scientist from JPL (Jet Propulsion Laboratory) in California, who had these words to say:

"A heavenly body as large as Jupiter, and (which is; sic) part of this solar system, has been found in the direction of the constellation of Orion by an orbiting telescope."

Further evidence of the existence of Planet X has been found by the research of Tom van Flandern, and also by John Anderson, a planetary scientist at JPL. The general public has not been made aware of the existence of this heavenly body, however, for more or less obvious reasons: to avoid mass panic and societal chaos.

One may wonder why humanity hasn't heard of this mysterious planet, estimated to be four times as massive as Earth, yet twenty times as dense--in effect, a "brown star," a dwarf star which either never ignited or which has all but burned out. But humanity has known about this object, and many ancient cultures recorded it, tracked it, and anticipated its cycles of approach, for it cuts through the solar system as part of its regular but unusual orbit, every 3600 years or so, creating interplanetary chaos as it goes.

To the ancient Sumerians this celestial body was known as the 12th planet (they counted the sun and moon as planets) and as NIBIRU, the Planet of the Crossing, the astronomical symbol of which was a CROSS. NIBIRU is also phonetically similar to the original name of the Hebrews,

"Hibiru" or "Habiru." The Babylonians, Assyrians, and other Mesopotamians called the dark planet Marduk, The King of The Heavens. Hebrews astronomers of antiquity referred to it as the Winged Globe, the king of celestial objects which sojourned among the stars themselves. To the Greeks, as well as to some modern researchers, it was and is known as Nemesis, the Bringer of Destruction. According to Sumerian calculations, it had an approximate orbit of 3600 years, which was called a SHAR. The Vedas tell us that Hindu astronomers named the same passage of time, marked by periodic destruction, as Treta Yuga, a period of 3600 years, and the devastation caused by the planet's passing was known as a Kali Yuga. According to their calculations, we are at the end of a Treta Yuga cycle now.

Some estimates place the return of NIBIRU or Planet X somewhere perhaps between 2003 and 2013, or later. What does this "planetary upheaval," raging oceans and earthquakes, along with the ancient symbol of the CROSS as its planetary emblem, signify?

Luke 21: 27-28 states:

27 And then they shall see the Son of man (Jesus Christ) coming in a cloud with power and great glory.

28 And when these things come to pass, then look up, and lift up your heads; for your redemption (salvation) draws nigh.

To clarify scriptures which have long-been considered "symbolic" or "esoteric," and point out the scientific and physical validity and meaning of them, is not always an easy thing. According to Revelation as well as other books, for the coming span of time which New Agers and their unseen manipulators see as the "New Age," there will be quite a different era upon the Earth, an entire millennium of peace, prosperity, health, and moral righteousness (according to Judeo-Christian standards). This will be due to the "benevolent dictatorship," the theocracy of Him who "rode upon the white horse," and who has gathered up all NON-HUMAN enemies and imprisoned them for the duration BENEATH THE EARTH:

Revelation 20:2-3

2 And he laid hold on the dragon, that old serpent, which is the Devil, and Satan, and bound him a thousand years,

3 And cast him into the bottomless pit, and shut him up, and set a seal (lock) upon him, that he should deceive the nations no more, till the thousand years be fulfilled: and after that he must be loosed a little season (a brief period of time, apparently for appeals, trial, judgment, and sentencing).

During this time when evil influences are removed, great advances in longevity and knowledge will be realized. Nature will be restored to its former health and glory upon the planet. A literal "spacestation-city" will remain in geosynchronous orbit above the earthly city of Jerusalem (Revelation Chapter 21) and people will pass back and forth between the "earthly" and "heavenly" Jerusalems. Interestingly, some parts of this "heavenly structure" are described as being transparent as glass, yet made of gold: "and the city was pure gold, like unto clear glass (Revelation 21:14)." It should be noted that the helmet visors of all Apollo astronauts were transparent, yet coated inside with a thin layer of pure gold, as this was determined to be the ONLY material which will allow the passage of visible light, yet deflect all harmful solar and cosmic radiation!

The reign of shadowy "unseen empires" is coming to a close. The members of humanity are faced with a choice: Allying themselves with a doomed kingdom of deception which has victimized us for thousands of years, or with a greater galactic kingdom, the "Kingdom of Heaven," a spiritual and physical reality which will soon dominate this planet for good.

The "coming of the saucers" indeed heralds a "new age," but not that expected by many. It heralds the return and reign of Jesus Christ, the Messiah of Mankind, Lord of a galactic or universal empire, along with His heavenly armies--and the end of an age of deception and terror which has plagued mankind for far too long.

According to Milton, the fallen angels who followed Satan into rebellion where given serpentine, reptilian forms/natures as punishment.

Illustration: Gustave Dore

(See Color Inserts)

The War in Heaven. Anthropromorphic angels, or galactic war and flying disks the outcome is the same. The King wins

Illustration: Gustave Dore

The "reptilian" or fallen-angelic corruption of humanity, with the Tree of Knowledge —sexual and genetic knowledge— as a major motive behind the interaction?

Illustration: William Blake

A meeting in the underworld, complete with demons...er, fallen angels....uh, reptilian humanoids....er, Nephilim....Titans? Jotuns?

Illustration: Gustave Dore

BIBLIOGRAPHY
non-fiction sources:

1. Above Top Secret: The Worldwide UFO Coverup, by Timothy Good. William Morrow, 1988
2. Alien Contact: Top Secret UFO Files Revealed, by Timothy Good. William Morrow, 1993
3. American Elves, An Encyclopedia of Little People from the Lore of 380 Ethnic Groups of the Western Hemisphere, by John E. Roth. MacFarland Books, 1997
4. American Indian Myths and Mysteries, by Vincent H. Gaddis. Indian Head Books, New York
5. An Encyclopedia of Fairies, by Katherine Briggs. Pantheon Books, 1976
6. Architects of the Underworld, by Bruce Rux. Frog Ltd., 1996
7. The Arabian Nights' Entertainments, or the Book of a Thousand Nights and a Night, by Sir Richard Francis Burton, Avenel Books.
8. The Bell Witch of Tennessee, by Charles Bailey Bell, 1934. Facsimile Edition by Charles Elder, Publisher, 1972
9. Beowulf, translated by Burton Raffel. New American Library, Mentor Books, 1963
10. Bhagavad-Gita As It Is, Abridged Edition, A. C. Bhaktivedanta Swami Prabhupada, The Bhaktivedanta Book Trust, 1972
11. The Bible Code, by Michael Drosnin, 1999, Simon & Schuster, 1998
12. Bishinik Magazine, That Old Time Choctaw Religion, by Len Green, June 1979
13. Bishinik Magazine, Origins of the Choctaw People Retold from Old Legends, by Len Green, August 1980
14. Casebook on the Men in Black, by Jim Keith. Illuminet Press, 1997
15. Celtic Myth and Legend, Poetry and Romance, by Charles Squire. Bell Publishing Co., 1979

16. Chupacabras and Other Mysteries, by Scott Corrales. Greenleaf Publications, 1997
17. The Clear Mirror : A Traditional Account of Tibet's Golden Age : Sakyapa Sonam Gyaltsen's Clear Mirror on Royal Genealogy by Bsod-Nams-Rgyal-Mtshan, Lama Choedak Yuthok (Translator), Delek cho Shastri, Sakyapa Sonam Gyaltsen, McComas Taylor. Snow Lion Publications, 1996
18. The Controllers, by Commander X, Abelard Productions, Inc.
19. Disneyland of the Gods, by John A. Keel. Illuminet Press, 1995
20. The Field Guide to Bigfoot, Yeti, and Other Mystery Primates Worldwide. Loren Coleman and Patrick Huyghe, Avon Books, 1999
21. Flying Serpents and Dragons, by R.A. Boulay, The Book Tree, 1997, 1998
22. Folktales of Norway, Edited by Reidar Christiansen, Translated by Pat Shaw Iversen. University of Chicago Press, 1968
23. The Forgotten Books of Eden: Lost Books of the Old Testament, Edited by Rutherford H. Platt, Jr., J. Alden Brett. Gramercy Books, Random House, New York 1980
24. Gods and Myths of the Viking Age, by H. R. Ellis Davidson. Bell Publishing Co., 1981
25. The Great Orm of Loch Ness, by F.W. Holiday. W.W. Norton and Co., 1968
26. Hollow Planets, by Jan Lamprecht (ISBN: 0-620-21963-7), 1999
27. The Infamous Bell Witch of Tennessee, by Charles Edwin Price. The Overmountain Press, 1994
28. Irish Folktales, edited by Henry Glassie. Pantheon Books, 1985
29. Japanese Ghosts and Demons, edited by Stephen Addiss. 1985, George Braziller, Inc.
30. Japanese Tales, Edited and Compiled by Royall Tyler, Pantheon Books, 1987
31. The Kalevala by Elias Lonnrot, translated by Keith Bosley, Oxford University Press, 1989

32. Lake Monster Traditions: A Cross-Cultural Analysis, by Michel Meurger and Claude Gagnon. Fortean Tomes, 1988
33. Lost Continents and the Hollow Earth, by David Hatcher Childress and Richard S. Shaver. Adventures Unlimited Press, 1999
34. MAYA: The Riddle and Rediscovery of a Lost Civilization, by Charles Gallenkamp. Viking Penquin, 1985
35. The Mothman Prophecies, by John A. Keel. Illuminet Press, 1991
36. Myth and Mankind: Epics of Early Civilization: Myths of the Ancient Near East, Time-Life Books, 1998
37. Myth and Mankind: Epics of Early Civilization: Land of the Dragon, Time-Life Books, 1999
38. Myth and Mankind: Gods of Sun and Sacrifice: Aztec & Maya Myth, Time-Life Books, 1997
39. New Larousse Encyclopedia of Mythology, Hamlyn Publishing Group Ltd., London-New York-Sidney-Toronto. Sections referenced: Assyro-Babylonian, Egyptian, Celtic, Indian, Chinese, Japanese.
40. The Norse Myths, Kevin Crossley-Holland. Pantheon Books, 1980
41. Operation Trojan Horse, by John A. Keel. G.P. Putnam & Sons, 1970
42. Our Haunted Planet, by John A. Keel. Galde Press, 1999
43. Passing Strange: True Tales of New England Hauntings and Horrors, by Joseph A. Citro. Houghton Mifflin Co., 1997
44. Peer Gynt : A Dramatic Poem (Oxford World's Classics), by Henrik Johan Ibsen
45. Popol Vuh, Translated by Dennis Tedlock. Simon and Schuster, 1985
46. The Psychic Sasquatch and their UFO Connection, by Jack Lapseritis. Blue Water Publishing, 1998
47. The Ramayana, by Valmiki, translated by William Buck. New American Library, 1978
48. The Samurai: A Military History, Stephen R. Turnbull. Macmillan Publishing Co., 1977

49. Sasquatch/Bigfoot, by Don Hunter and Rene Dahinden. Revised Edition, McClelland & Stewart, 1993
50. Scandinavian Folk Belief and Legend, Reimund Kvideland, Henning K. Sehmsdorf, Eds. University of Minnesota Press, 1988.
51. The Secret Commonwealth, by Robert Kirk (1691). Published by Stirling, Eneas Mackay,1933, and edited by Andrew Lang (third edition).
52. The Secret World, by Ray Palmer (with Richard S. Shaver). Amherst Press, 1975
53. The Story of Sigurd (Volsunga Saga), Translated from the Icelandic by Eirikr Magnusson and William Morris, From A Treasury of Fantasy, Avenel Books, 1981
54. Superstitions of the Highlands and Islands of Scotland, by John Campbell
55. Subterranean Worlds, by Walter Kafton-Minkel. Loompanics Unlimited, 1989
56. Subterranean Worlds Inside Earth, by Timothy Green Beckley. Inner Light Productions, 1992
57. Unexplained! by Jerome Clark. Visible Ink Press, 1999
58. The Wilderness Hunter, by Theodore (Teddy) Roosevelt, 1892

BIBLIOGRAPHY of Fiction Sources:

All Fiction Sources are referenced within the body of the treatise.

IMAGES

- Illustrations by Donn P. Crane and Willy Pogany are from MY BOOK HOUSE, 1932.
- Illustrations by H.J. Ford and Lancelot Steed are taken from the folktale compilations of Andrew Lang:
- The Red Fairy Book, Longmans, Green, and Co., London, 1890
- The Green Fairy Book,, Longmans, Green, and Co., London, 1892
- The Yellow Fairy Book, Longmans, Green, and Co., London, 1894
- The Pink Fairy Book, Longmans, Green, and Co., London, 1897
- The Crimson Fairy Book, Longmans, Green, and Co., London, 1903
- The Brown Fairy Book, Longmans, Green, and Co., London, 1904
- The Lilac Fairy Book, Longmans, Green, and Co., London, 1910

Special thanks to HawkSword for providing the original editions for scanning purposes, as well as the scans of Ukiyo-e artwork by Utagawa Kuniyoshi. Special thanks to Adobe Photoshop for its restorative powers!

Special thanks the Dean de Luca (Dharmapada Dasa) for information on the Bhagavat Purana and the undersea kingdom mentioned therein.

Special thanks to all who provided image files of an archeological, cultural, historical, and cartographic nature.

ABOUT THE AUTHOR

Wm. Michael Mott is the Creative Director for a high-performance software company. He is also a freelance artist and writer, and writes both fiction and non-fiction. He has worked as an artist/designer for Fortune 500 companies, for an NSF Engineering Research Center, and for a variety of freelance clients such as book and magazine publishers.

His artwork has appeared in many publications, such as Computer Graphics World Magazine, Computer Artist, IEEE Computer, IEEE Computer Graphics & Applications, PHOTO/Electronic Imaging, DRAGON Magazine, and others. He's created artwork and graphic design for mass-market book covers, posters, brochures, packaging, CD-ROM covers, art collections, and digital/web-based media. Mr. Mott won several design awards, from regional Advertising Federation awards for printed material, to awards for web site graphics and design. His artwork has been featured in the exhibition "In Dreams Awake: Art of Fantasy" at the Olympia and York Gallery, 1988; at the 1987 World Fantasy Con, Con*stellation, and others.

His fantasy novels, "Pulsifer: A Fable," and "Land of Ice, A Velvet Knife," are published in electronic form by Softbook Press Inc. (http://www.softbook.com, now Gemstar), and a series of illustrated short stories began in 2000, as chapbooks from Undaunted Press (www.undauntedpress.com).

He is currently researching Fortean and paranormal topics. His research in this area has appeared in magazines such as Fate, Nexus, and World Explorer.

E-mail contact: mottimorph@earthlink.net

Galllery website: http://www.asfa-art.org/gallery/mott.html

Shaver Rock Images

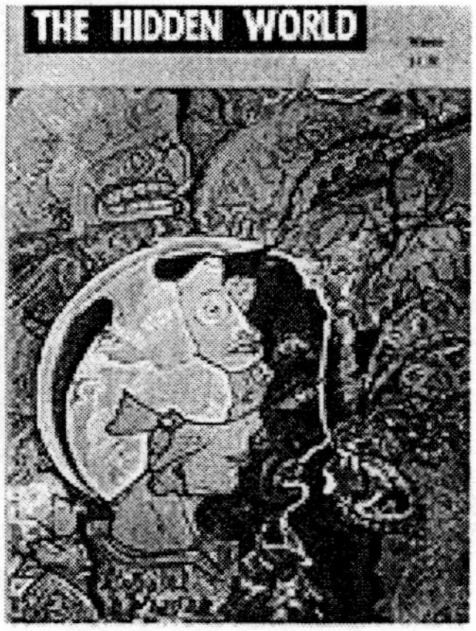

Just a few of the many magazines to feature the "true" stories and commentaries of Richard S. Shaver
(See Color Inserts)

Caverns, Cauldrons and Concealed Creatures

Some of the following Shaver rock images, and the letter to Ray Palmer have NEVER BEEN PUBLISHED BEFORE, to my knowledge. The letter was sent from Shaver to Palmer in 1972. They are, as Shaver tried to say (but he lacked the terminology), holographic images preserved in crystallized form, which he revealed by cutting with a rock saw. He would then shoot light through them with a projector, and make prints of all sizes. The color images and the mermaid image were featured by Shaver and Ray Palmer in The Secret World.

The Angry Twins

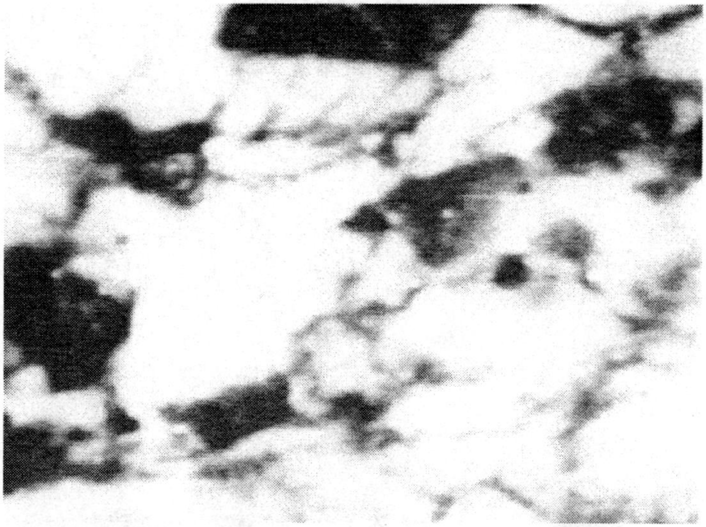

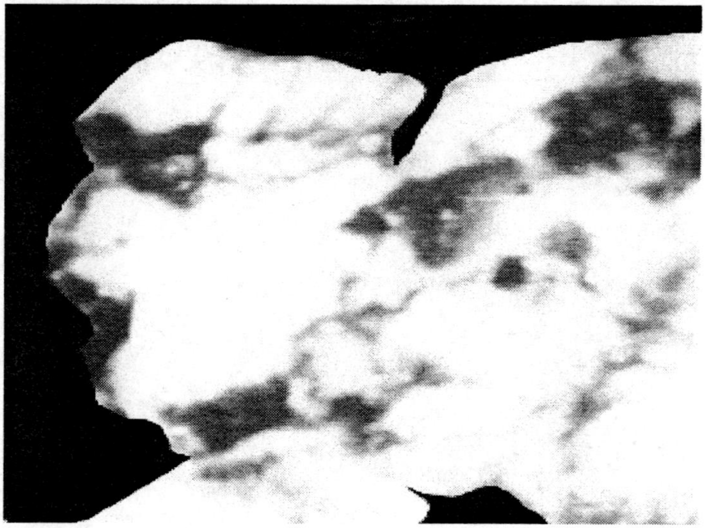

The Beauty Queen

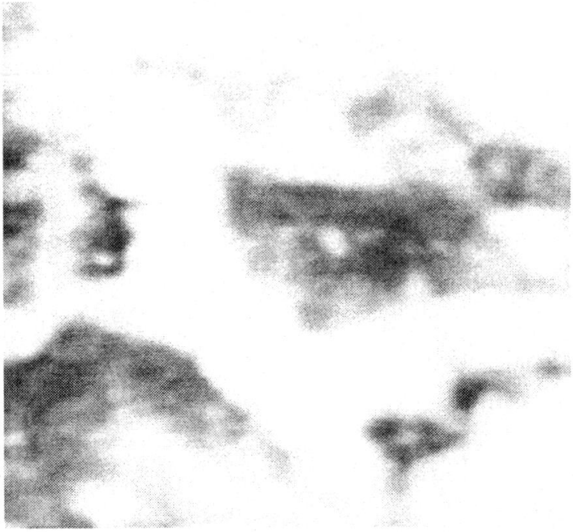

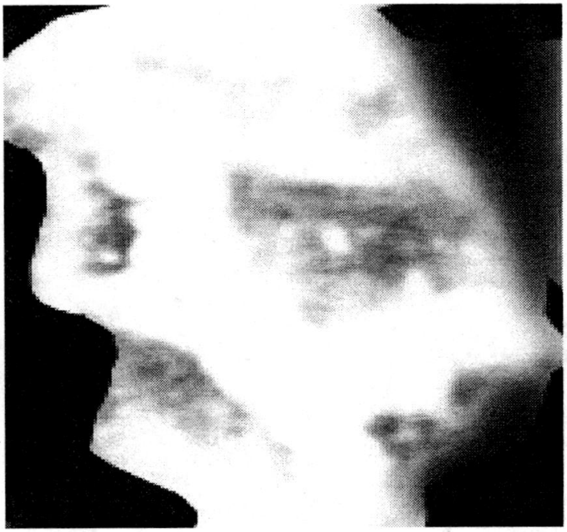

The Incan King

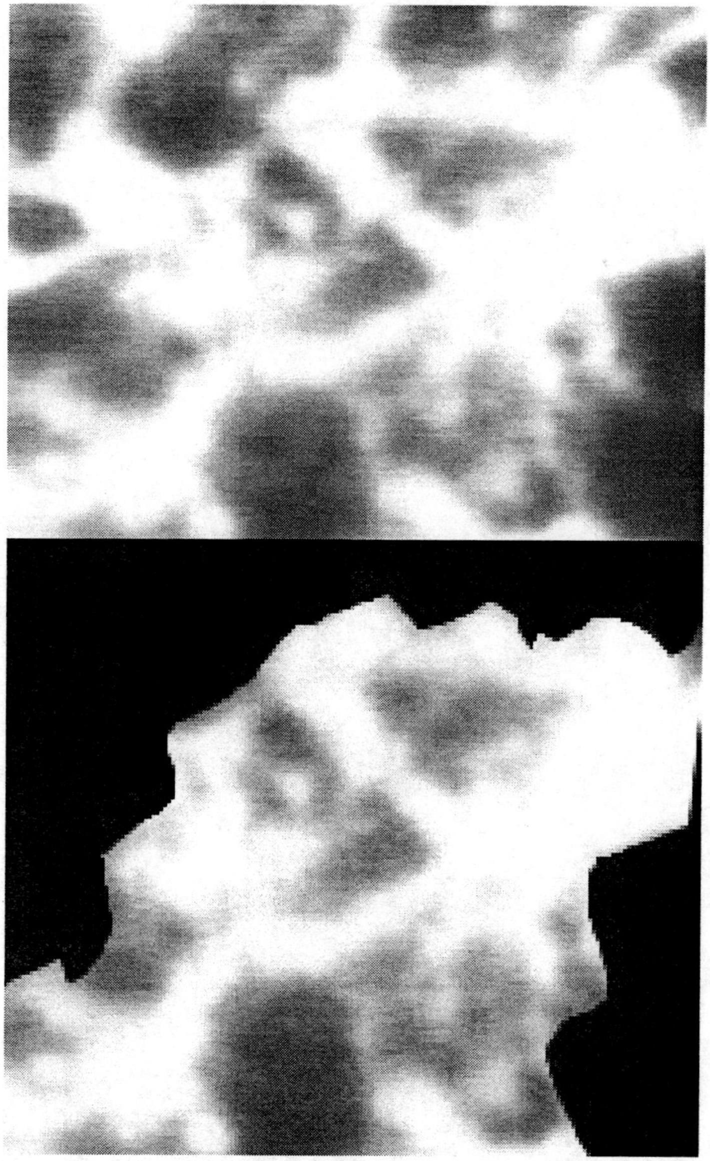

The Wizard

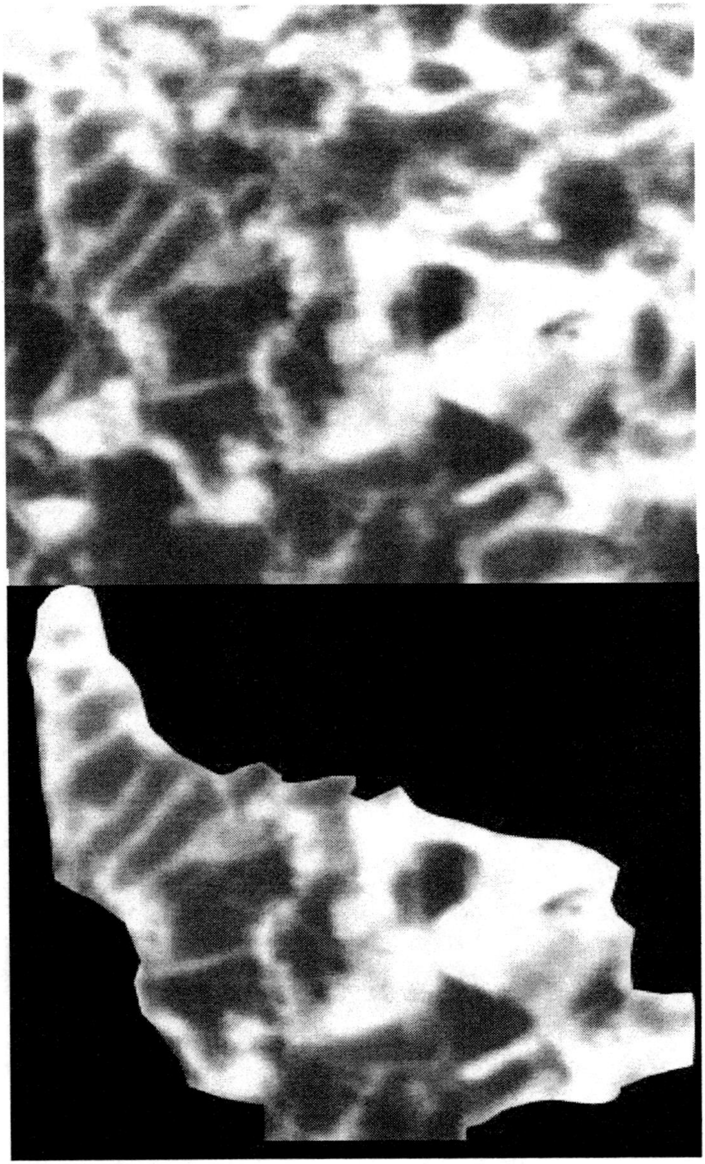

The Whisperers

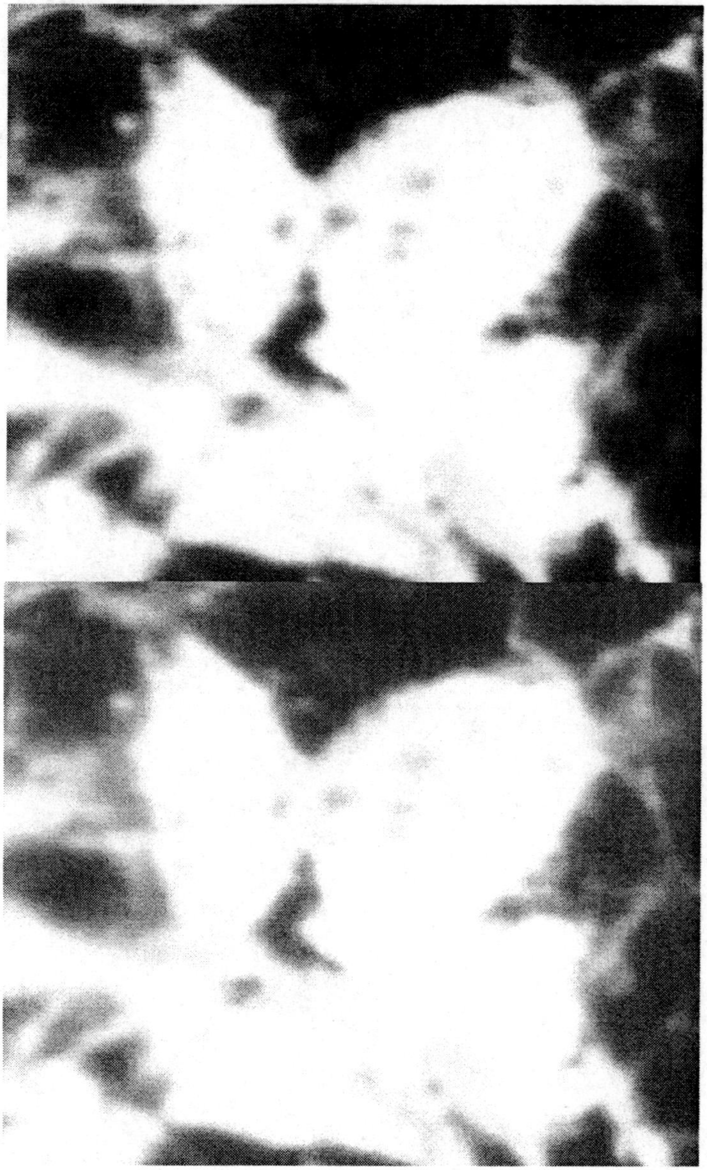

The Little Demon

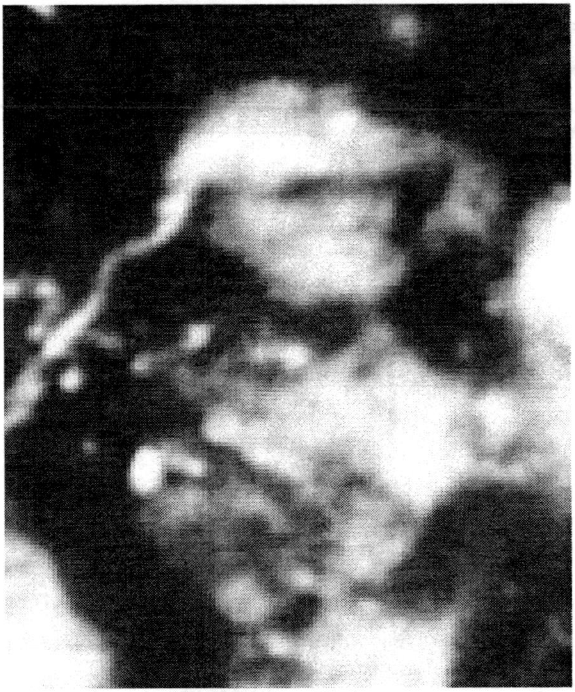

The Face Cluster

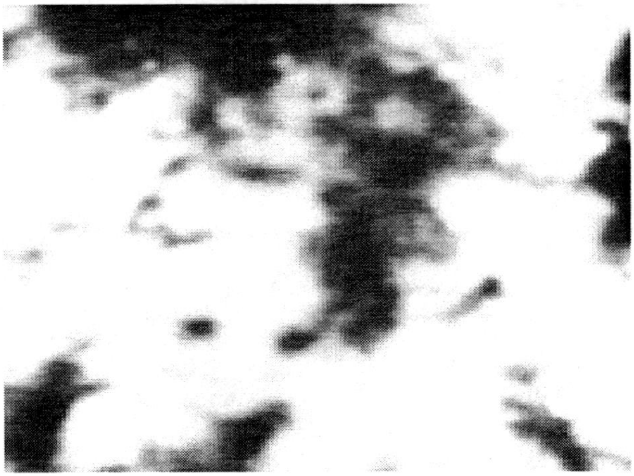

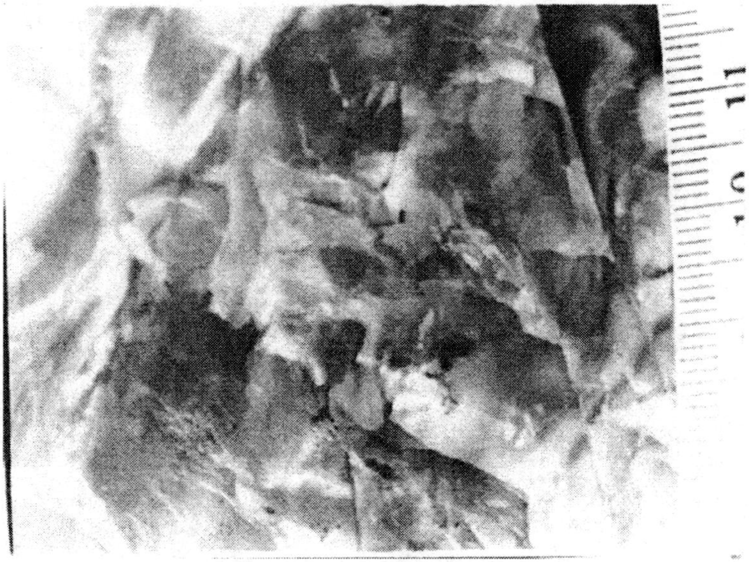

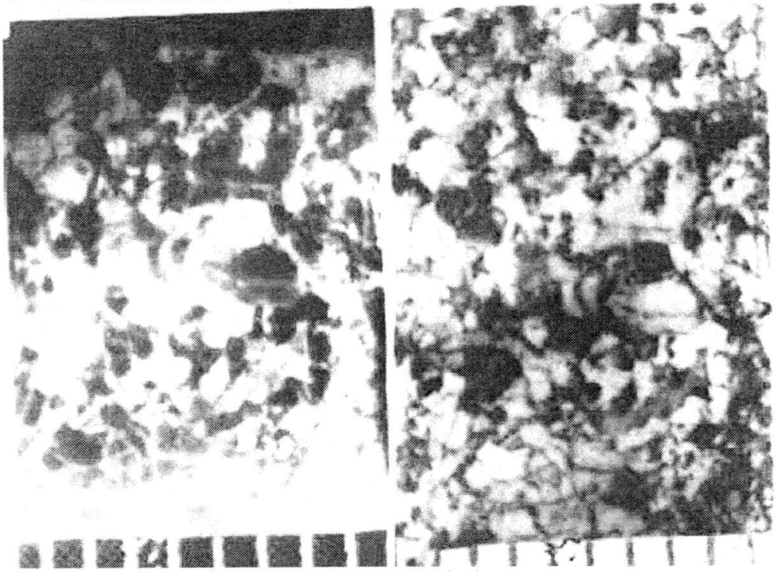

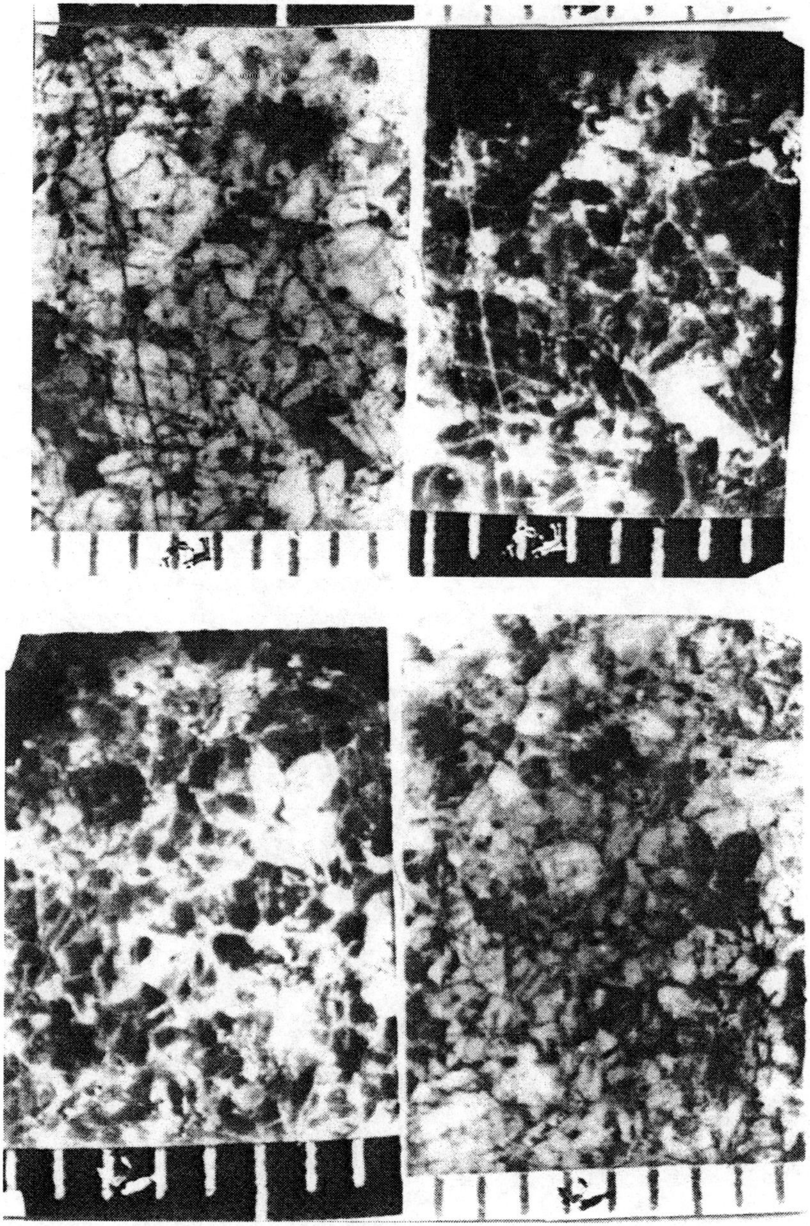

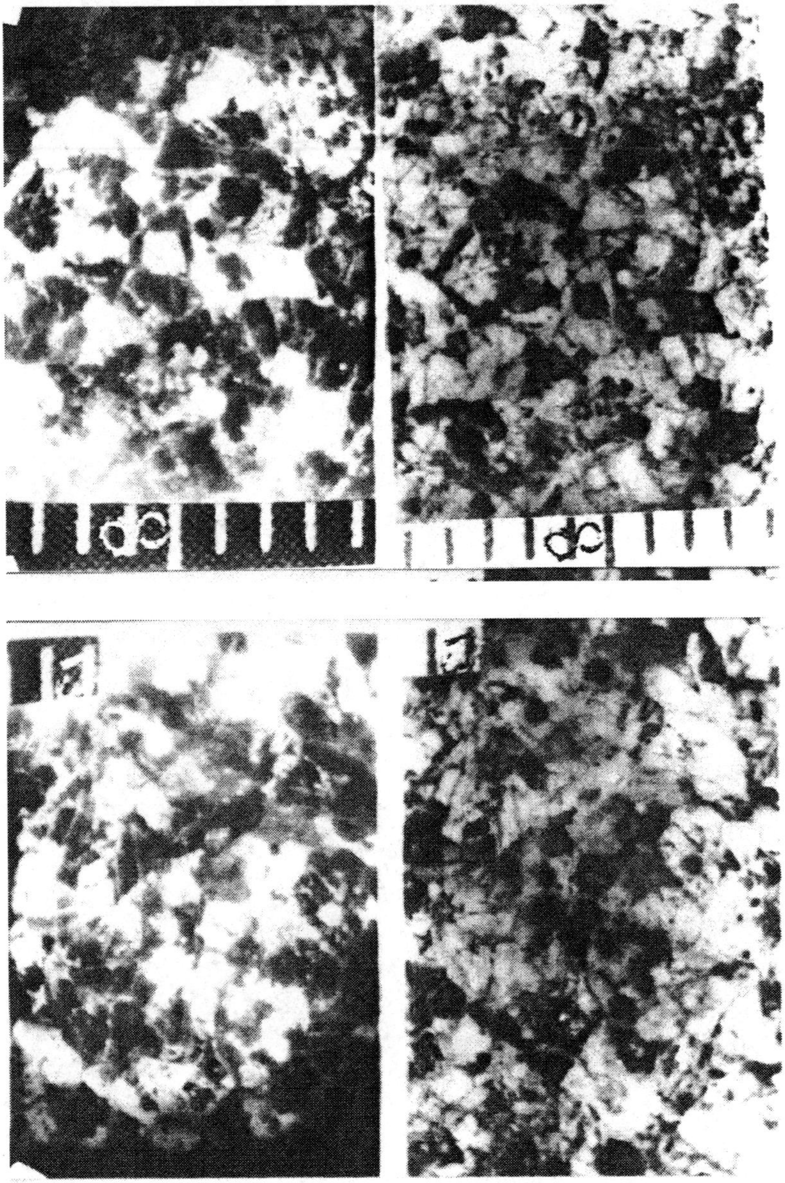

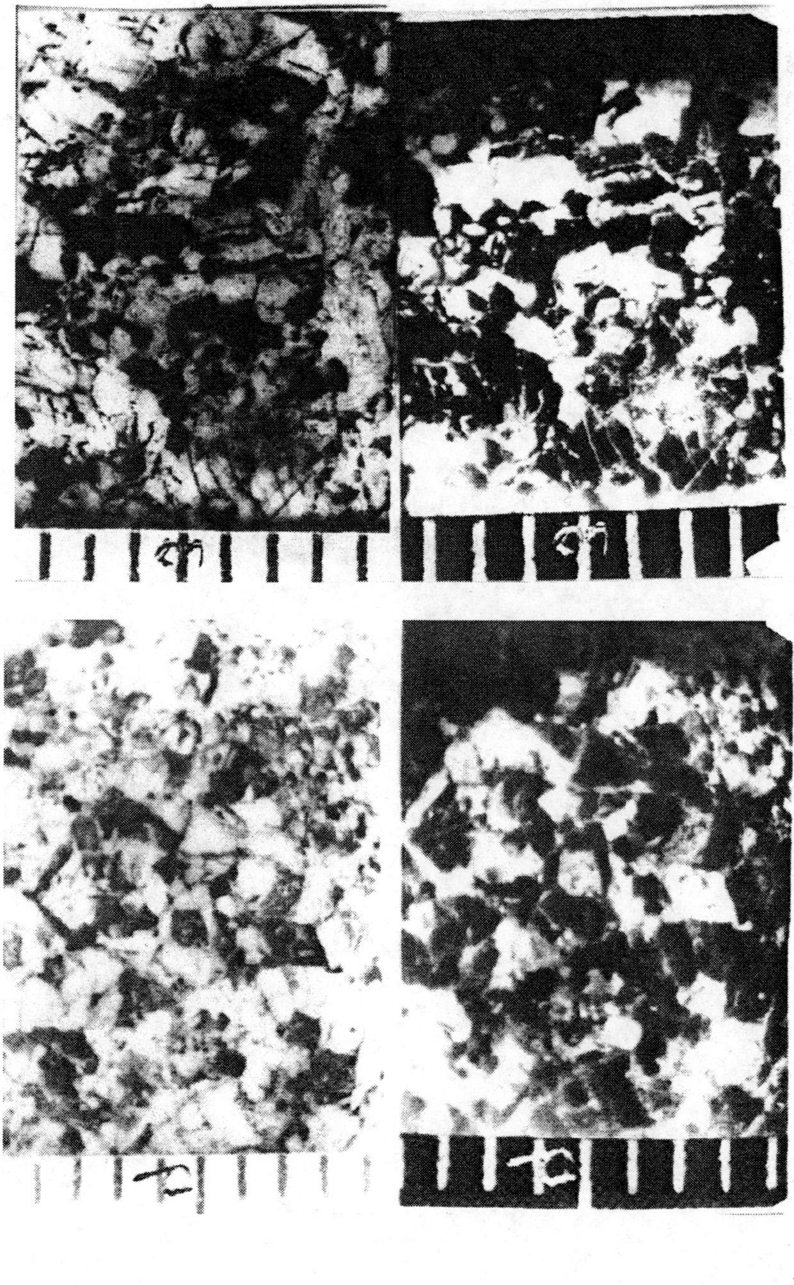

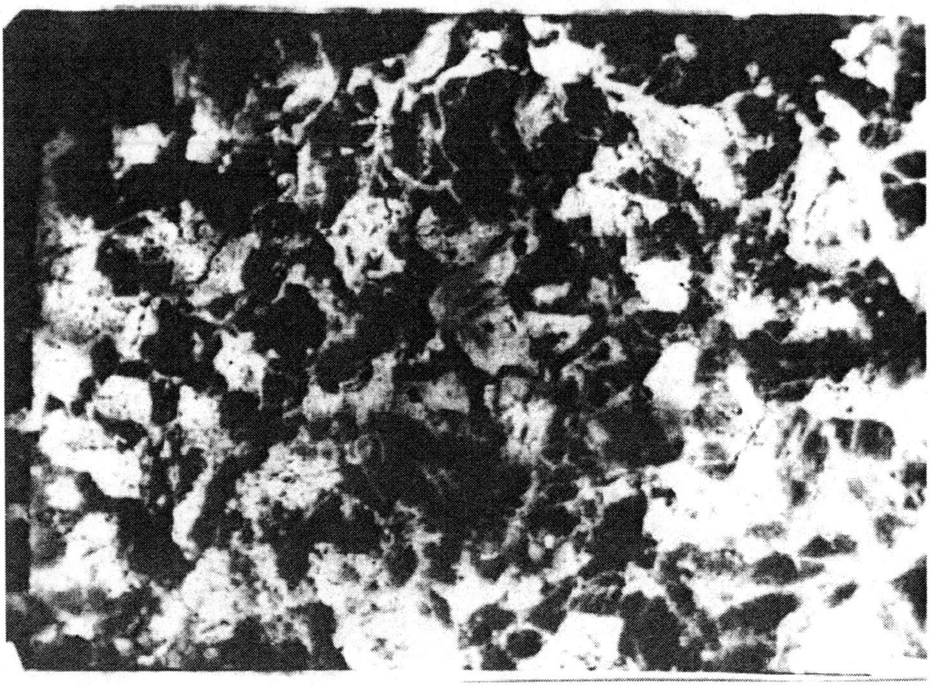

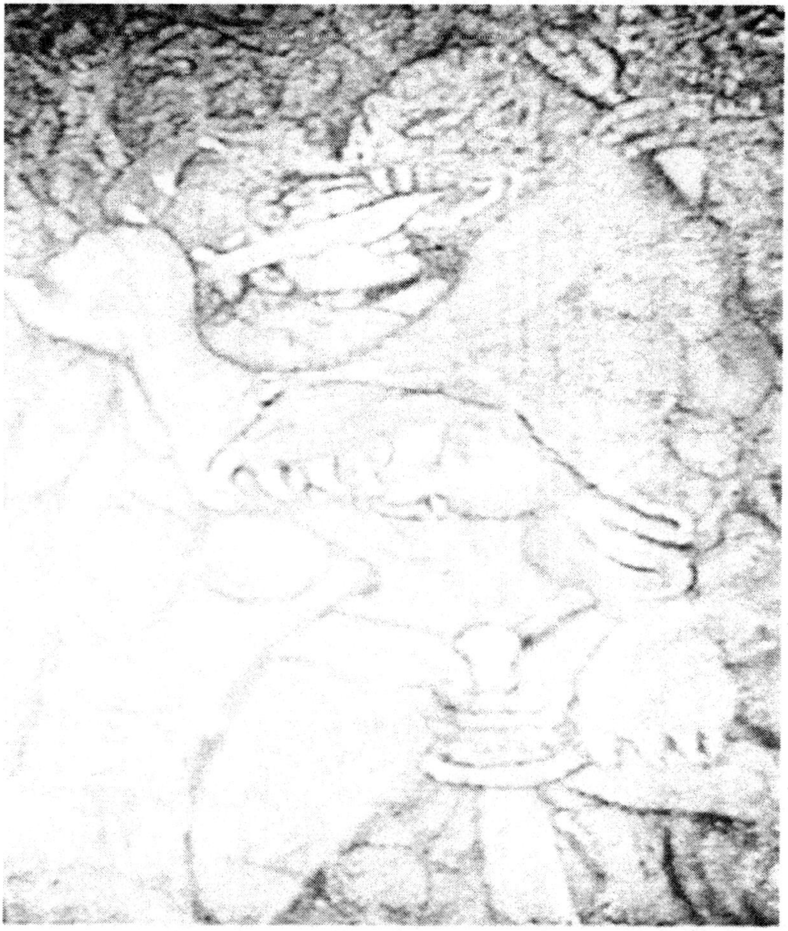

One of Richard Shaver's "Picture Rocks," cut with a rock-saw, and revealing human women fighting a reptilian humanoid similar in form to "El Chupacabras."

(See Color Inserts)

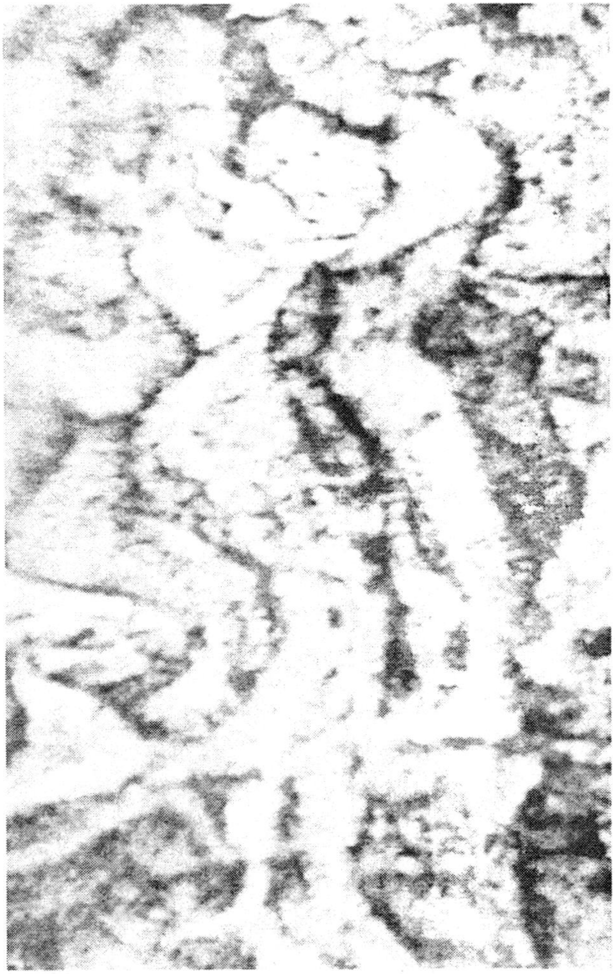

One of Richard Shaver's "picture rocks," depicting a mermaid in the act of presenting a hybrid child to a human father. Another, larger, matronly mermaid figure is poised behind the new mother. The infant looks human enough.

Shaver's rock images sometimes depicted merfolk, ranging from diminutive to gigantic, and aquatic creatures like dolphins.

speaks
with
forked
tongue
photo

large profile
covers
height of
print..

all large pic
tures are made
up of smaller

every face of
white has a
black to contrast
it ..and vice
versa

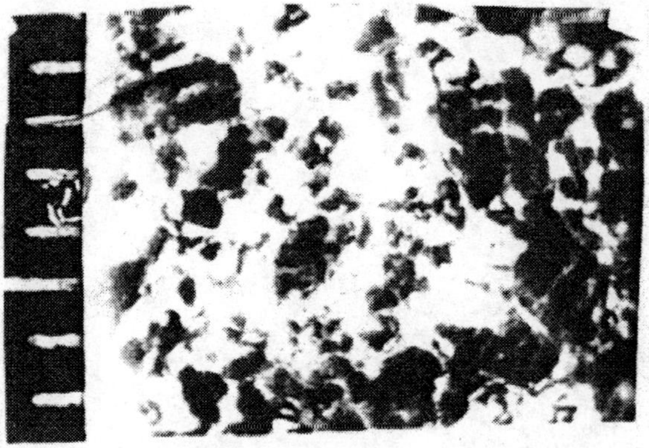

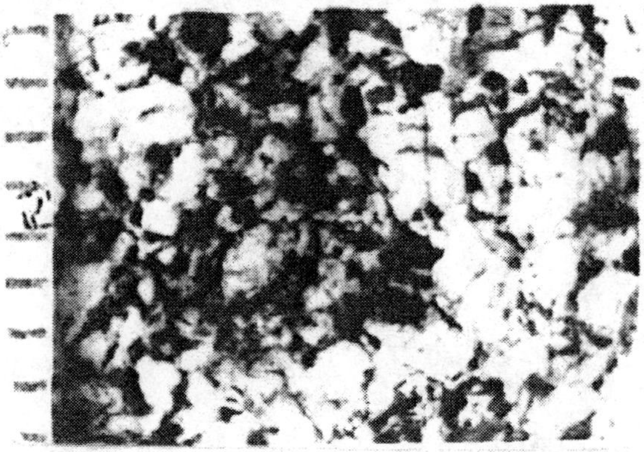

Rep..if you think my prints are bad or obscure or puzzling
why then take a real good look at a page full of Picasso prints
they are far far more o bscure and puzzling, and worse art by far...
and whats more..these are infinitely less childish in what message does
get across

One way to display these prints is to print a Picasso print of the wilder
kind (aren't theyall) and comment that Snover rock photos are more
intriguing and more thought provoking and even less puzzling by far

Shaver sometimes projected light through water-thin slices of his "rocks," using a projector. He would then create images on chemically-treated canvas of the things he believed he had found, often in slices of rock of miniscule size, and then he added color.

In this picture, two of his titan-like "Atlans" are shown, filled with joy over their genetic prosperity and bounty. This fascination with procreation and genetic diversity/ success is by now well-established. According to Shaver, the descendants of these "Atlans" and "Titans" went underground (or were they driven there, as were the Nephilim of old?) in order to escape harmful cosmic and solar radiation. Over the ages they degenerated into a variety of stunted and misshapen forms.

(See Color Inserts)

THE
BOOKS
OF
ENOCH
PUBLISHED BY
TGS SERVICES

The Book
of Enoch
Translated from the Ethiopic
Translated by George H. Schodde - 1882

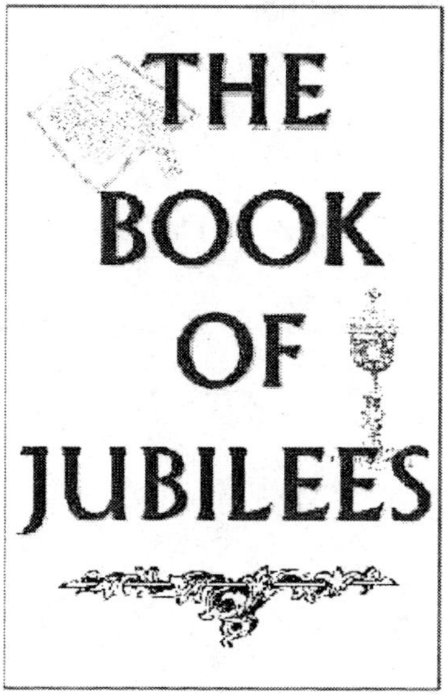
THE
BOOK
OF
JUBILEES

THE
BOOK
OF
JASHER

TGS Publishers
22241 Pinedale Lane
Frankston, Texas 75763

HiddenMysteries.com
903-876-3256

Photo Image © 2000, Wm. Michael Mott

Illustration for Dante's Inferno, by Donn P. Crane

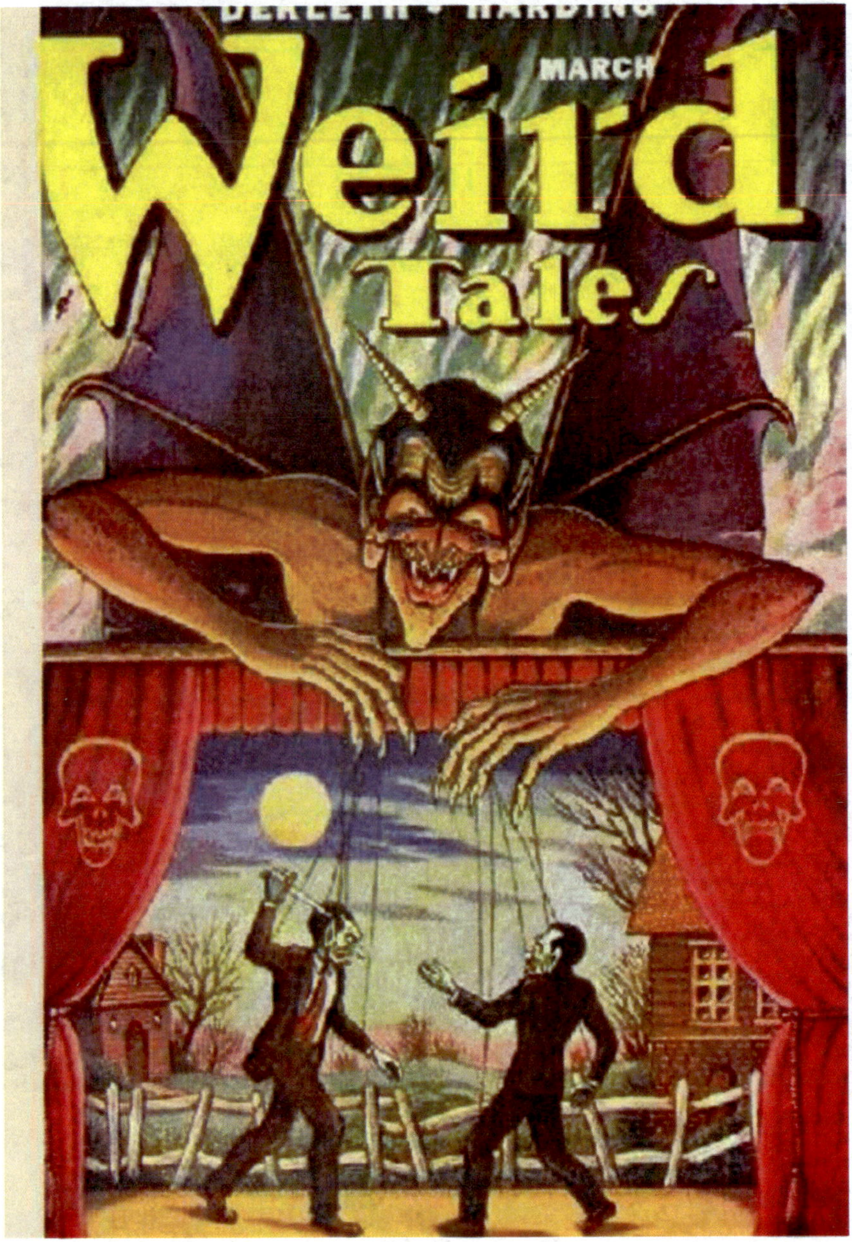
MARCH
Weird Tales

THE MAN WHO CRIED "WOLF!" by ROBERT BLOCH
MAY
Weird Tales
15¢
The dazzling danger of . . .
"The Shining Land"
by Edmond Hamilton

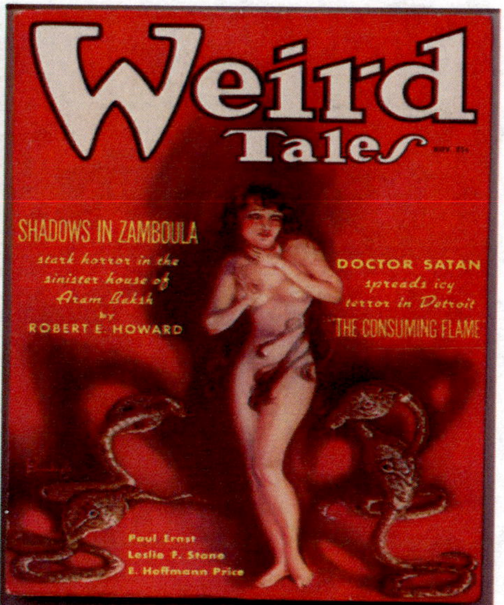
Weird Tales
SHADOWS IN ZAMBOULA
stark horror in the sinister house of Aram Baksh
by
ROBERT E. HOWARD
DOCTOR SATAN
spreads icy terror in Detroit
"THE CONSUMING FLAME"
Paul Ernst
Leslie F. Stone
E. Hoffmann Price

Weird Tales
AUG.
25c
THE
DEVIL
IN IRON
WEIRD ADVENTURE IN AN AMAZING ISLAND CITY OF GREEN STONE
By ROBERT E. HOWARD
C. L. MOORE
HUGH B. CAVE
FRANCIS FLAGG
ARLTON EADIE
F. B. LONG, JR.

Otis Adelbert Kline in this issue
Weird Tales
The Unique Magazine
A STARTLING
VAMPIRE STORY
By KIRK MASHBURN
ROBERT E. HOWARD
CLARK ASHTON SMITH
W. ELWYN BACKUS
PAUL ERNST
ALEXANDRE DUMAS

Caverns, Cauldrons and Concealed Creatures

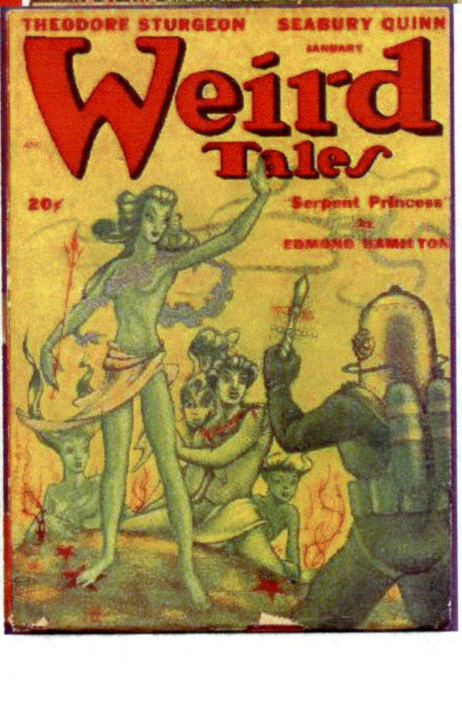

Caverns, Cauldrons and Concealed Creatures

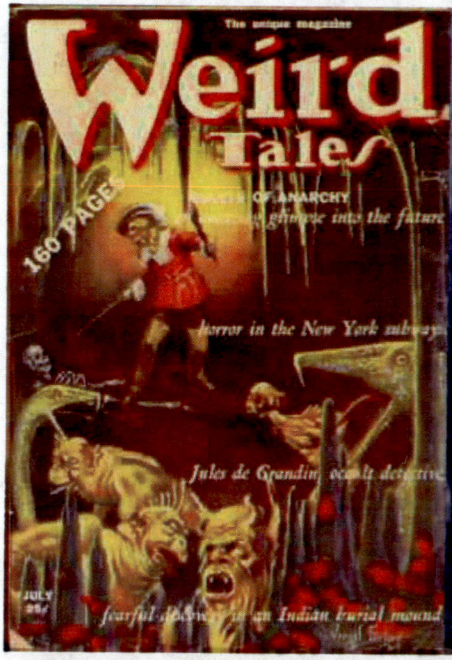

Caverns, Cauldrons and Concealed Creatures

One of Richard Shaver's "Picture Rocks," cut with a rock-saw,and revealing human women fighting a reptilian humanoid similar in form to "El Chupacabras."

© William Michael Mott

© William Michael Mott

© William Michael Mott

© William Michael Mott

Tengu, Japanese bird-headed (reptilian) humanoids.

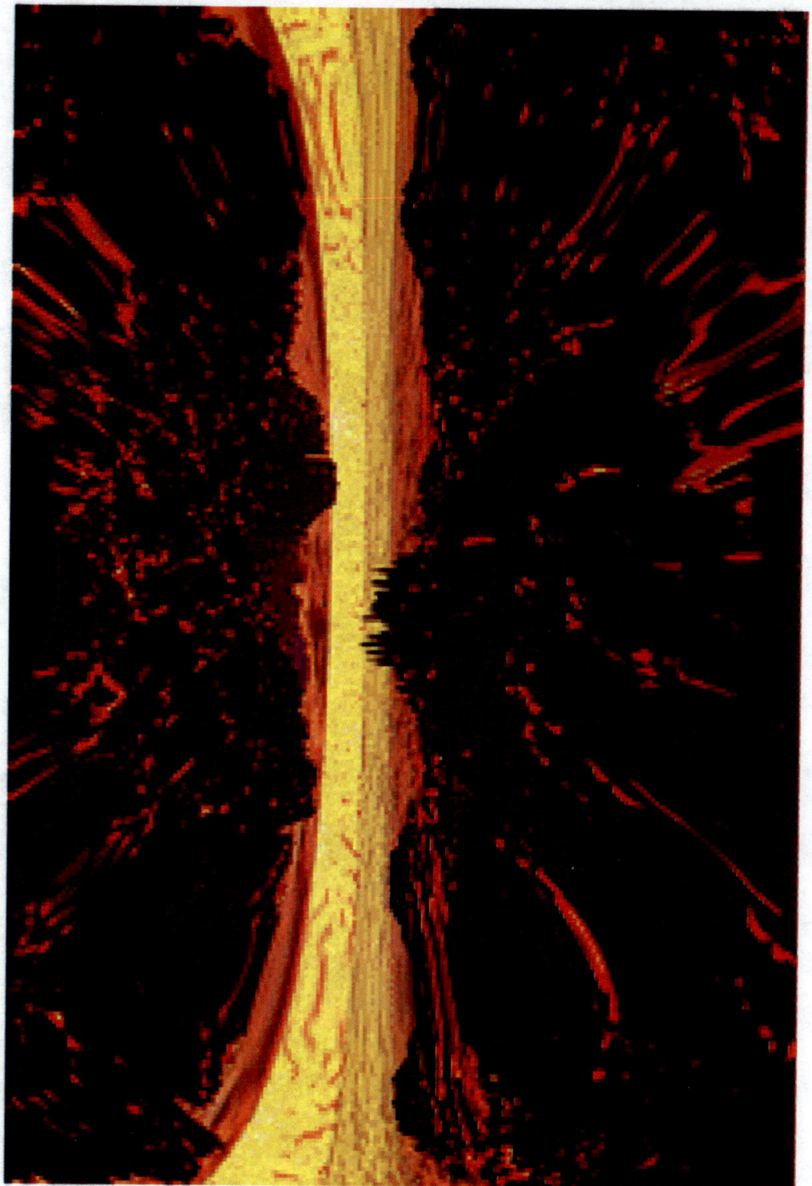

© William Michael Mott

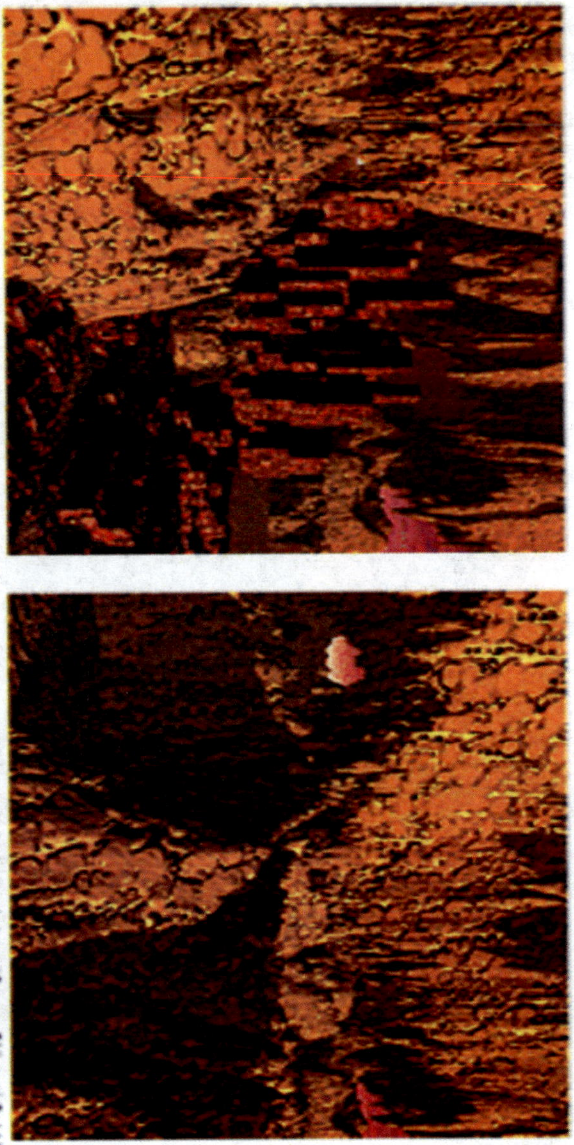

According to Aztec origin myth, they originated from Chicomoztoc (the seven caves), from which came the legend of the "Seven Cities of Gold."

Perseus rescuing the "virgin" Andromeda, and her genetic potential, from the "dragon" or leviathan of the sea.

Illustration: Willy Pogány

Two views of the incubus or hag-ridden phenomenon, by Fuseli.

The author took this photo beside a cave entrance in the mountains of Northern Alabama in 1992. The entrance is small, located in a high bluff and only accessible by way of a dangerous ledge. Not until after the film was developed did the visage of a peak-capped, scowling, dwarf-like figure become evident, apparently made from different colorations of rock and lichen. Note the details of beard, lips, teeth nose (with nostrils) cheekbones, eyes and brows.

Whatever it is, it is an amazing simulacrum of a garbed humanoid being

**Beowulf and Grendel,
By Donn P. Crane**

According to Milton, the fallen angels who followed Satan into rebellion where given serpentine, reptilian forms/natures as punishment.

Illustration: Gustave Dore

Some examples of the Egyptian underworld deities, Anubis and Ammut.

Beowulf confronts Grendel's mother in her lair, as drawn by Donn P. Crane

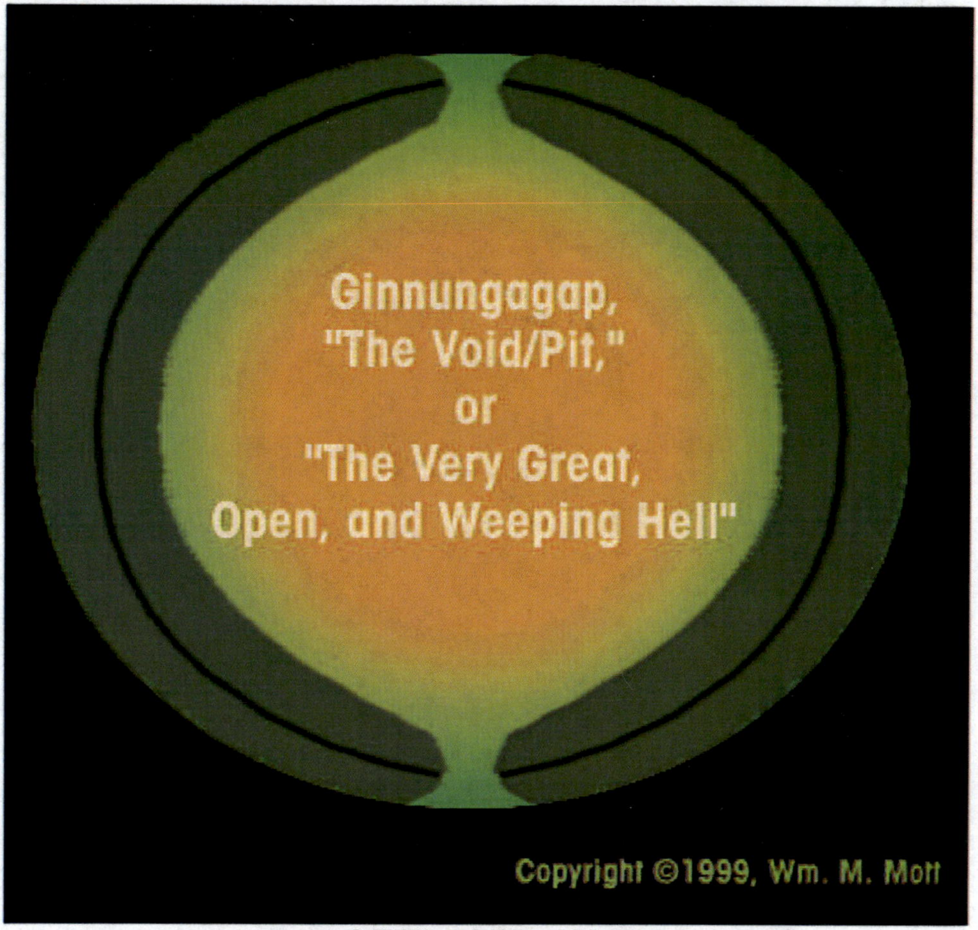

© William Michael Mott

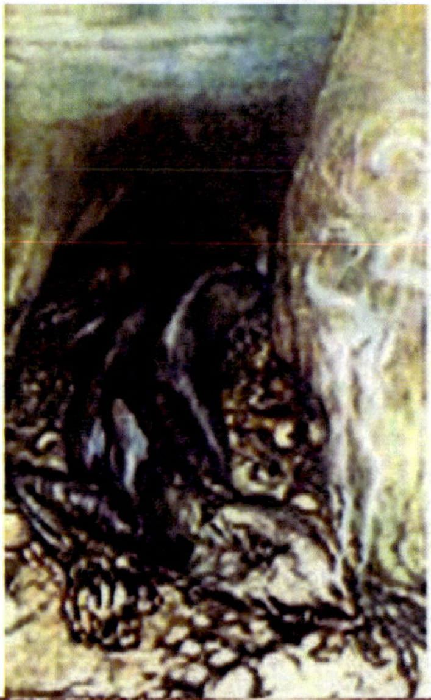

© William Michael Mott

© William Michael Mott

© William Michael Mott

Above: Photos taken from Apollo 16, showing an unusual weather formation near the "anomalous area." The photos were taken hours apart, and show a strange zone with a clearly-demarcated edge, and a thin inner coating of clouds. While not completely stationary, the area pictured raises the question: What polar conditions created such a meteorological oddity at the top of the globe?

Left and Bottom: Photo taken from 22,300 miles away, by NASA's Application Technology Satellite III, showing the entire planet. As the enlargement at bottom shows, the anomaly at the pole is clearly seen, including the inner sweep of cloud cover as it pours over the "lip." Given the relative size of the Earth to the opening, it is much higher, deeper, and is also wider.

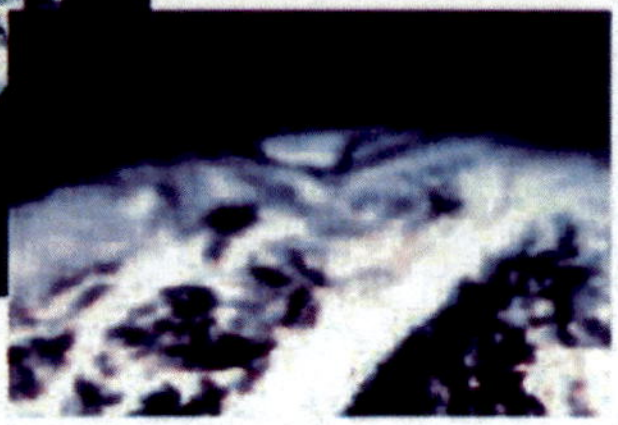

Some say this NASA photo show the 'lip' of an entrance to an unknown region of the Earth

© William Michael Mott

© William Michael Mott

© William Michael Mott

Mott '98

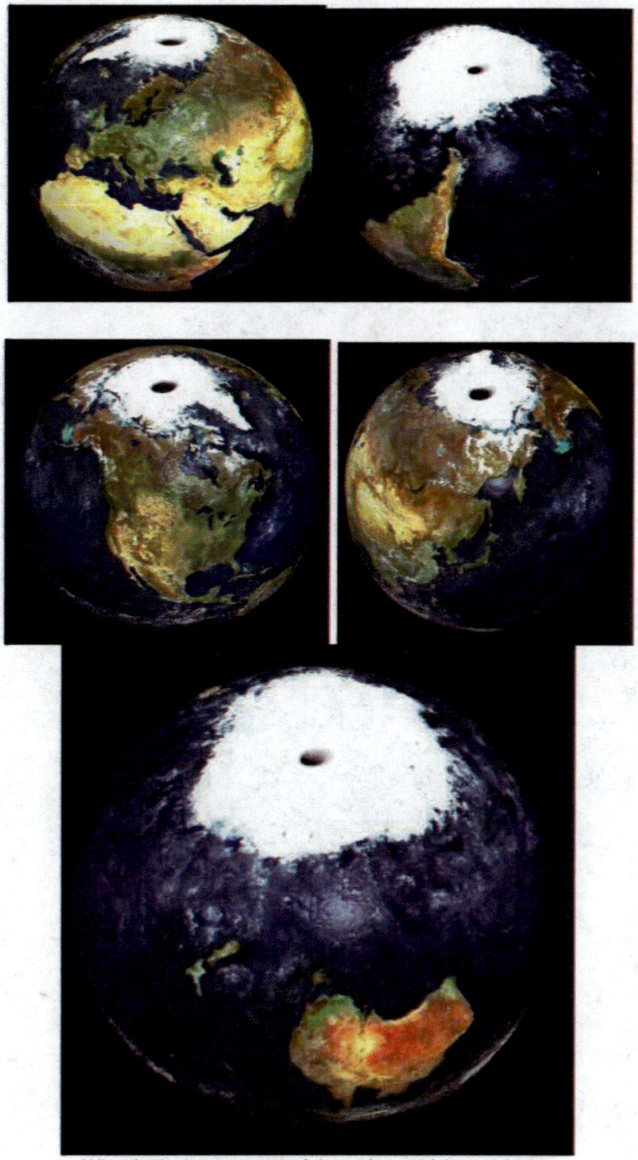

What is the true nature of the poles, which are almost always concealed by thick clouds or fog, and is there a link to legend of a vast underworld?

© William Michael Mott

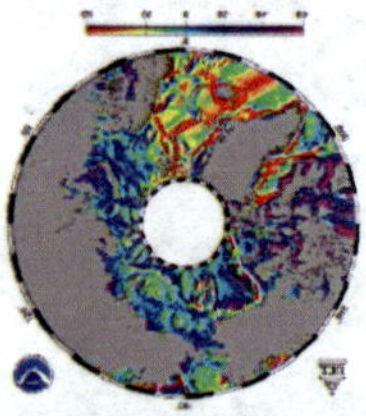

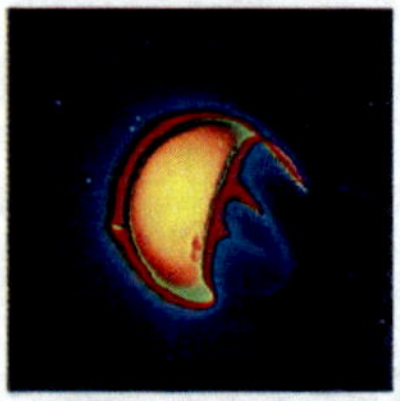

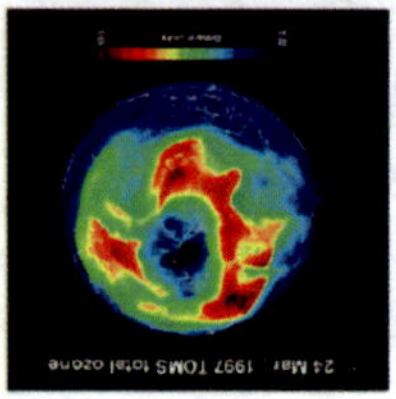

In centuries past, traveling "morality plays" would pass through towns and countryside, the finale featuring the "Hell Mouth:" a devouring reptilian dragon or devil's head, into which actors were forced by others in "hairy humanoid" or demonic garb.

Illustration: Donn P. Crane

A Sasquatch or Bigfoot, drawn by a ten-year old child.

The Nanih Waiya Cave Mound © Wm.Michael Mott 2002

Goya captured the horrors of cannibalism in his painting depicting Saturn, devouring his offspring.

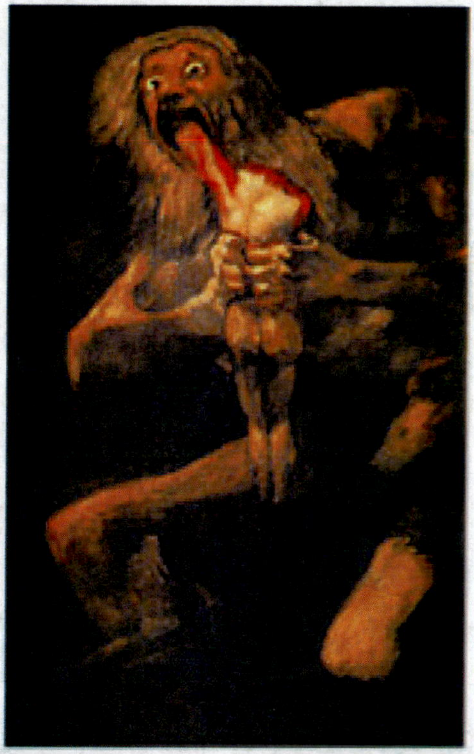

The 20th Century saw a new spin on the "Giants in the Earth:" The "Shaver Mystery."

The Western view of
the underworld, or Hell, has much in
common with traditions from around the
world: the abode of maleficent beings,
demons, and monsters, and the place of
torment for the human unrighteous.
In western religious thought, this view
has changed little from Medieval
and Renaissance times, until the present.

According to Western tradition, the
fallen angels are imprisoned within
the Earth, along with the nefilim, a later
band of transgressors who lusted
after human women. This theme is
paralleled from the imprisonment of
the Titans and Jotuns, to the modern
confabulations of Richard Shaver.

Scenes from The Damned and the Elect,
and The Elect Being Called to Paradise
and The Damned Being Plunged into Hell
(1499-1502),
by Luca Signorelli

and

Fall of the Rebel Angels (1620), top center,
by by Pieter Pauwel Ruben.

© William Michael Mott

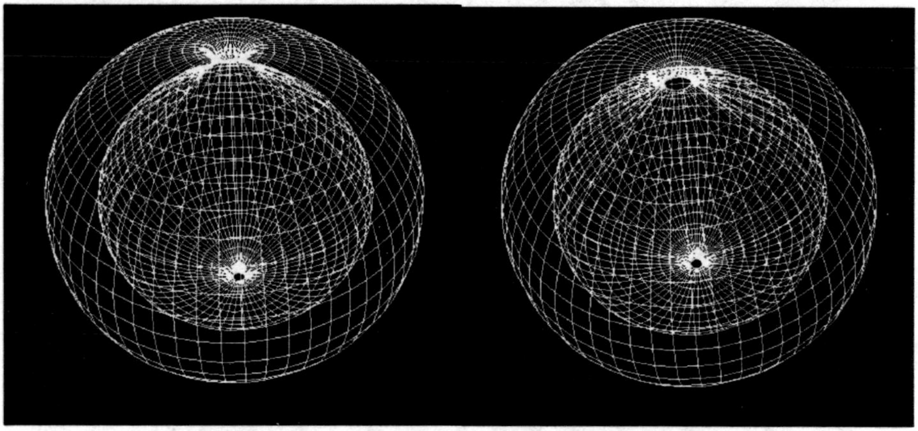

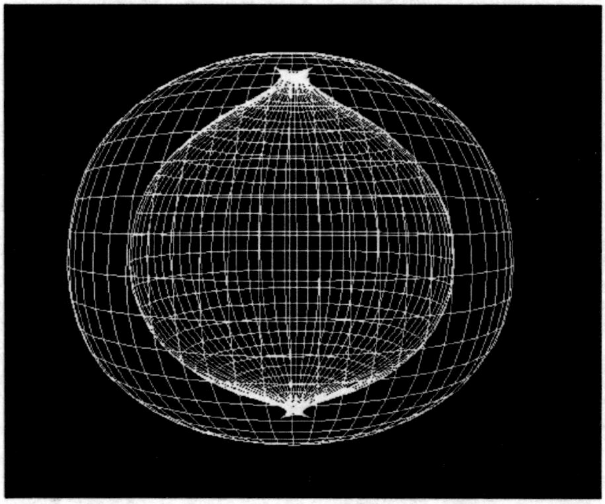